AF322739

Genome Informatics 2007

GENOME INFORMATICS SERIES (GIS)
ISSN: 0919-9454

The Genome Informatics Series publishes peer-reviewed papers presented at the International Conference on Genome Informatics (GIW) and some conferences on bioinformatics. The Genome Informatics Series is indexed in MEDLINE.

No.	Title	Year	ISBN Cl./Pa.
1	Genome Informatics Workshop I	1990	(in Japanese)
2	Genome Informatics Workshop II	1991	(in Japanese)
3	Genome Informatics Workshop III	1992	(in Japanese)
4	Genome Informatics Workshop IV	1993	4-946443-20-7
5	Genome Informatics Workshop 1994	1994	4-946443-24-X
6	Genome Informatics Workship 1995	1995	4-946443-33-9
7	Genome Informatics 1996	1996	4-946443-37-1
8	Genome Informatics 1997	1997	4-946443-47-9
9	Genome Informatics 1998	1998	4-946443-52-5
10	Genome Informatics 1999	1999	4-946443-59-2
11	Genome Informatics 2000	2000	4-946443-65-7
12	Genome Informatics 2001	2001	4-946443-72-X
13	Genome Informatics 2002	2002	4-946443-79-7
14	Genome Informatics 2003	2003	4-946443-82-7
15	Genome Informatics 2004 Vol. 15, No. 1	2004	4-946443-88-6
16	Genome Informatics 2004 Vol. 15, No. 2	2004	4-946443-91-6
17	Genome Informatics 2005 Vol. 16, No. 1	2005	4-946443-93-2
18	Genome Informatics 2005 Vol. 16, No. 2	2005	4-946443-96-7
19	Genome Informatics 2006 Vol. 17, No. 1	2006	4-946443-97-5
20	Genome Informatics 2006 Vol. 17, No. 2	2006	4-946443-99-1
21	Genome Informatics 2007 Vol. 18	2007	Forthcoming
22	Genome Informatics 2007 Vol. 19	2007	978-1-86094-984-5

Genome Informatics Series Vol. 19   ISSN: 0919-9454

Genome Informatics 2007

PROCEEDINGS OF THE 18TH INTERNATIONAL CONFERENCE

BIOPOLIS, SINGAPORE   3 – 5 DECEMBER 2007

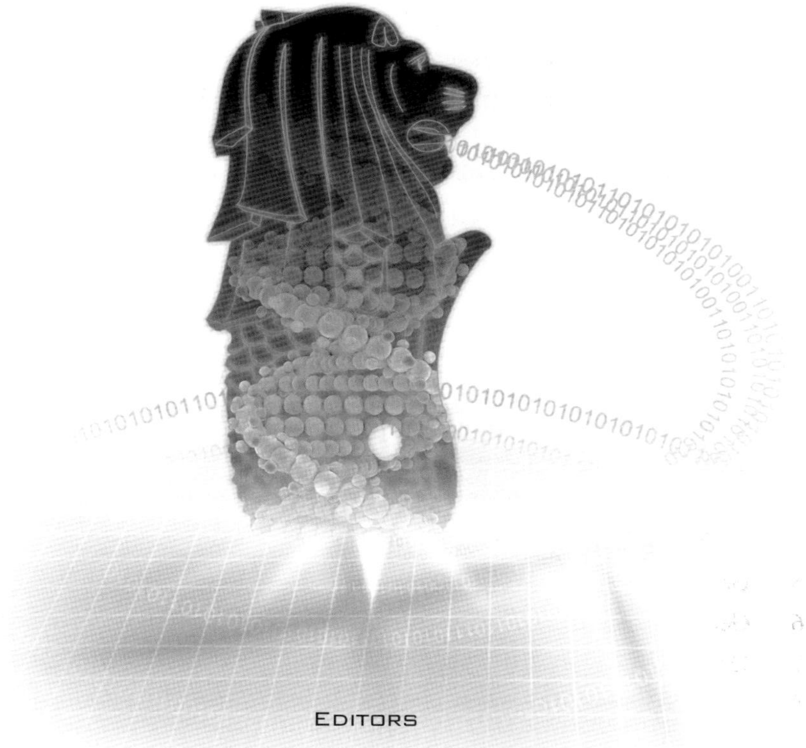

EDITORS

SEE-KIONG NG
Institute for Infocomm Research, Singapore

HIROSHI MAMITSUKA
Kyoto University, Japan

LIMSOON WONG
National University of Singapore, Singapore

Imperial College Press

Published by

Imperial College Press
57 Shelton Street
Covent Garden
London WC2H 9HE

Distributed by

World Scientific Publishing Co. Pte. Ltd.
5 Toh Tuck Link, Singapore 596224
USA office: 27 Warren Street, Suite 401-402, Hackensack, NJ 07601
UK office: 57 Shelton Street, Covent Garden, London WC2H 9HE

British Library Cataloguing-in-Publication Data
A catalogue record for this book is available from the British Library.

GENOME INFORMATICS 2007
Proceedings of the 18th International Conference

Copyright © 2007 by Imperial College Press

ISBN-13 978-1-86094-984-5
ISBN-10 1-86094-984-3

Printed in Singapore by Mainland Press Pte Ltd

CONTENTS

PREFACE

This book contains papers presented at the Eighteenth International Conference on Genome Informatics (GIW 2007) held in Biopolis, Singapore from December 3rd to 5th, 2007.

The GIW series provides an international forum for presentation and discussion of original research papers on all aspects of bioinformatics, computational biology and systems biology. Its scope includes biological sequence analysis, protein folding prediction, gene regulatory network, clustering algorithms, comparative genomics, and text mining. GIW has a history of 18 years and it is likely the longest running international bioinformatics conference. The first GIW was held at Kikai Shinko Kaikan, Tokyo during December 3-4, 1990 as an open workshop just before the Japanese Human Genome Project started in 1991. GIW 2007 was the first time that the conference was held outside of Japan.

The program committee of GIW 2007 received a total of 56 submissions from authors around the world (18 countries in total). Each submitted paper was reviewed by at least three members of the program committee. Based on their reports, 16 papers were accepted (29%) for presentation at the conference. These 16 papers appear in this book and are indexed in Medline. In addition, this book contains abstracts from the five invited speakers: Frank Eisenhaber, Bioinformatics Institute (Singapore), Sir David Lane, Institute of Molecular and Cell Biology (Singapore), Hanah Margalit, The Hebrew University of Jerusalem (Israel), Lawrence Stanton, Genome Institute of Singapore (Singapore), and Michael Zhang, Cold Spring Harbor Laboratory (USA).

The electronic versions of all these papers in this issue are also publicly available from the website of the Japanese Society for Bioinformatics (JSBi) (`http://www.jsbi.org/journal.html`).

See-Kiong Ng
Hiroshi Mamitsuka
GIW 2007 Program Committee Co-Chairs

Limsoon Wong
GIW 2007 Conference Chair

ACKNOWLEDGMENTS

First of all, we would like to thank all the authors for their effort in preparing their manuscripts. We also appreciate the great efforts made by the program committee members to ensure the high quality of the papers accepted—there were much spirited discussions during the discussion phase of the reviewing process, reflecting the dedication of the program committee members in reviewing the manuscripts despite their busy schedules. We also acknowledge the School of Computing and the Bioinformatics Programme at the National University of Singapore for hosting GIW 2007, the assistance from the local organizing committee members for arranging the conference venue, as well as the support of the Bioinformatics Institute and the Institute for Infocomm Research, Agency for Science, Technology and Research (A*STAR) of Singapore. We are also grateful for the generous sponsorships from World Scientific Publishing Company, Taylor & Francis Asia Pacific, and John Wiley & Sons Asia. Finally, we would like to give special thanks to those who presented papers or posters at GIW 2007, and those attended the conference. GIW 2007 would not be a complete success without their enthusiastic participation.

Organizers

Supporting Organizers

Gold Sponsor

Silver Sponsors

PROGRAM COMMITTEE

See-Kiong Ng	– Institute for Infocomm Research, Singapore; Co-Chair
Hiroshi Mamitsuka	– Kyoto University, Japan; Co-Chair
Gary Bader	– University of Toronto, Canada
Vladimir Bajic	– SANBI, South Africa
Christopher Baker	– Institute for Infocomm Research, Singapore
Ziv Bar-Joseph	– Carnegie Mellon University, USA
Guillaume Bourque	– Genome Institute of Singapore, Singapore
Jung-Hsien Chiang	– National Cheng Kung University, Taiwan
Francis YL Chin	– University of Hong Kong, Hong Kong
Peter Clote	– Boston College, USA
Chris HQ Ding	– Lawrence Berkeley National Laboratory, USA
Andreas Dress	– CAS-MPG Partner Institute of Computational Biology, China
Tamás Horváth	– University of Bonn and Fraunhofer IAIS, Germany
Wen-Lian Hsu	– Academia Sinica,Taiwan
Xiaohua Tony Hu	– Drexel University, USA
Seiya Imoto	– University of Tokyo, Japan
Minoru Kanehisa	– Kyoto University, Japan
George Karypis	– University of Minnesota, USA
Edda Klipp	– Max Planck Institute for Molecular Genetics, Germany
Ernst-Walter Knapp	– Free University Berlin, Germany
Stefen Kramer	– Technische Universitat Munchen, Germany
Dong-Yup Lee	– Bioprocessing Institute & National University of Singapore, Singapore
Sang Yup Lee	– KAIST, Korea
Ming Li	– University of Waterloo, Canada
Kui Lin	– Beijing Normal University, China
Frédérique Lisacek	– Swiss Institute of Bioinformatics, Switzerland
Aleksandar Milosavljevic	– Baylor College of Medicine, USA
Satoru Miyano	– University of Tokyo,Japan
Bernard Moret	– Swiss Federal Institute of Technology, Switzerland
Shin-ichi Morishita	– University of Tokyo, Japan
Richard Mott	– Wellcome Trust Centre for Human Genetics, UK
William Stafford Noble	– University of Washington, USA
Laxmi Parida	– IBM T. J. Watson Research Center, USA

Ron Pinter   – Technion, Israel
Mark Ragan   – University of Queensland, Australia
Yasubumi Sakakibara   – Keio University, Japan
Christian Schönbach   – Nanyang Technological University, Singapore
Tetsuo Shibuya   – University of Tokyo,Japan
Wing Kin Sung   – National University of Singapore, Singapore
Koji Tsuda   – Max Planck Institute for Biological Cybernetics, Germany
Alfonso Valencia   – Universidad Autonoma, Spain
Gabriel Valiente   – Technical University of Catalonia, Spain
Chandra Verma   – Bioinformatics Institute, Singapore
Jean-Philippe Vert   – Ecole des Mines de Paris, France
Martin Vingron   – Max Planck Institute for Molecular Genetics, Germany
Lusheng Wang   – The City University of Hong Kong, Hong Kong
Edwin Wang   – National Research Council Biotechnology Research
                           Institute, Canada
Eric Xing   – Carnegie Melon University, USA
Ying Xu   – University of Georgia, USA
Gwan-Su Yi   – ICU, Korea
Mohammed J. Zaki   – Rensselaer Polytechnic Institute, USA

CO-REVIEWERS

Rezwan Ahmed	Ricardo Alberich	Cheong Xin Chan
Dongsheng Che	Phuongan Dam	Masashi Fujita
Clarie Gervais	Ilan Gronau	Jun-tao Guo
Masumi Itoh	Aditi Kanhere	Karin Sonja Kassahn
Chris Kauffman	Yong Lu	Jin Hwan Park
Yanjun Qi	Jian Qiu	Huzefa Rangwala
Kengo Sato	Michael Shmoish	Robert Thurman
Richard Tzong-Han Tsai	Katsuyuki Yugi	

POSTER COMMITTEE

Guillaume Bourque   – Genome Institute of Singapore, Singapore;
                                         Chair
Alan Christoffels   – Temasek Lifesciences Laboratory, Singapore
Radha Krishna Murthy Karuturi   – Genome Institute of Singapore, Singapore
Chee Keong Kwoh   – Nanyang Technological University, Singapore
Dong-Yup Lee   – National University of Singapore, Singapore
Xiaoli Li   – Institute for Infocomm Research, Singapore
Olivo Miotto   – Institute of Systems Science, Singapore
Sanjay Swarup   – National University of Singapore, Singapore
Chandra Verma   – Bioinformatics Institute, Singapore

STEERING COMMITTEE

Minoru Kanehisa	– Kyoto University, Japan
Satoru Miyano	– University of Tokyo, Japan
Toshihisa Takagi	– University of Tokyo, Japan
Limsoon Wong	– National University of Singapore, Singapore

CONFERENCE CHAIR

Limsoon Wong – National University of Singapore, Singapore

ORGANIZING COMMITTEE

Agnes Ang	– National University of Singapore, Singapore
Lay Khim Chng	– National University of Singapore, Singapore
Kwok Pui Choi	– National University of Singapore, Singapore
Alexia Leong	– National University of Singapore, Singapore
Hon Wai Leong	– National University of Singapore, Singapore
Wai Kin Leong	– National University of Singapore, Singapore
Lay Hoon Liow	– National University of Singapore, Singapore
Stefanie Ng	– National University of Singapore, Singapore
Wing-Kin Sung	– National University of Singapore, Singapore
Martti Tammi	– National University of Singapore, Singapore
Siang Yong Yap	– National University of Singapore, Singapore
Xin Chen	– Nanyang Technological University, Singapore
Chee Keong Kwoh	– Nanyang Technological University, Singapore
Guillaume Bourque	– Genome Institute of Singapore, Singapore
Alan Christoffels	– Temasek Lifesciences Laboratory, Singapore
Dong-Yup Lee	– Bioprocessing Technology Institute, Singapore
Gunaretnam Rajagopal	– BioInformatics Institute, Singapore

PART A
Full Papers

DETECTION OF MONOSACCHARIDE TYPES FROM COORDINATES

MASANORI ARITA[1,2,3]   TOSHIAKI TOKIMATSU[1]
arita@k.u-tokyo.ac.jp   tokimatsu@cb.k.u-tokyo.ac.jp

[1] *Department of Computational Biology, Graduate School of Frontier Sciences, University of Tokyo and PRESTO JST, 5-1-5 CB05 Kashiwanoha, Kashiwa, 277-8561 Japan*
[2] *Plant Science Center, RIKEN, 1-7-22 Suehiro-cho, Tsurumi, Yokohama, 230-0045 Japan*
[3] *Institute of Advanced Biosciences, Keio Univ., 14-1 Baba-cho, Tsuruoka, 997-0035 Japan*

Almost half of biological molecules (proteins and metabolites) are extrapolated as glycosylated within cells. Detection of glycosylation patterns and of attached sugar types is therefore an important step in future glycomics research. We present two algorithms to detect sugar types in Haworth projection, *i.e.*, from x-y coordinates. The algorithms were applied to the database of flavonoid and identified backbone-specific biases of sugar types and their conjugated positions. The algorithms contribute not only to bridge between polysaccharide databases and pathway databases, but also to detect structural errors in metabolic databases.

Keywords: algorithm, flavonoid, monosaccharide, stereochemistry

1.   Introduction

Glycosylation is a major post-transcriptional and post-translational modification for biological molecules. Two thirds of all proteins are extrapolated as glycosylated to become functional [2], and the same is true for many secondary metabolites. For example, among 6,850 flavonoid species structurally identified to date, as much as 50 % were found in glycosylated form [16]. Despite such universality, a common, computational notation for oligo- and polysaccharides that can be used for research articles and databases has been missing in glycomics. The situation clearly decelerated the development and integration of carbohydrate information resources when compared against other molecular information such as genome and protein sequences. A notable exception is the Glycan database, which uses a graphical layout for polysaccharide molecular structures [12].

Although the structure of carbohydrates is much more complicated than linear DNAs and protein sequences, linear systematic codes have been already proposed [3,5]. In time, a particular code would become the universal standard, and the next problem is how to translate currently available carbohydrate information in different formats into such a systematic code. Obviously manual conversion is time-consuming and error-prone, and there is an urgent demand for a computational solution. The purpose of this contribution is to provide an automated technique that can convert conventional descriptions into a systematic code, in preparation for a possible standard description of saccharide structures. The crucial step is the recognition of major drawing styles to describe the stereochemistry of cyclic monosaccharides; oligo- and polysaccharides are their repetitions and can be treated similarly. Moreover, less than 5-carbon or more than 6-carbon monosaccharides such as tetrose and heptose are rare as natural modifiers. For this reason we focus here on the recognition of 5- or 6-carbon cyclic monosaccharides. In drawing molecular structures, frequently used styles are the following: Mills depiction, Haworth projection, and its variation for three-dimensional view. The guidelines of each drawing style are formally recommended by IUPAC-IUBMB as Nomenclature of Carbohydrates 2-Carb-5 [9].

The paper is organized as follows. Since depiction scheme of sugar structures is not well known, we introduce the major drawing styles and type of sugars after this introduction. In Chapter 2, the formal detection scheme is presented. The procedure was implemented in Java and was tested on our Flavonoid databases [16]. The result is introduced in Chapter 3, followed by the conclusion and future work of this study.

1.1. *The Fischer projection*

The standard method in teaching stereochemistry of monosaccharides is the Fischer projection. It represents every stereocenter as a cross. The horizontal line represents bonds extending above the paper plane, and the vertical line represents bonds extending below the plane (Figure 1). The IUPAC numbering of 6 carbons in hexose starts from top (C-1) to bottom (C-6) direction. This projection is used in textbooks only, but we introduce it for easier understanding of the following other descriptions.

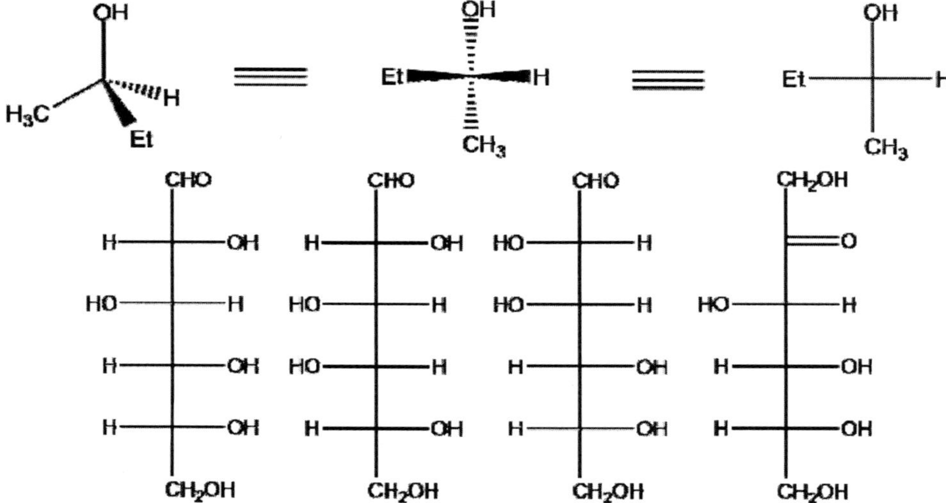

Figure 1. Fischer projection and four hexose examples. From the left, D-glucose, D-galactose, D-mannose and D-fructose. The topmost carbons are C-1, and bottommost are C-6. The figures are taken from http://www.metabolome.jp/doc/lectures/biochem/sugar/.

1.2. *The Mills depiction*

The Mills method is widely used in organic chemistry and molecular biology, including web-based databases such as the Kyoto Encyclopedia of Genes and Genomes database (KEGG) [11], PubChem database [15], and metabolomics web-services such as Biological Magnetic Resonance Data Bank (BMRB) [4]. In the method, the ring of monosaccharide is set on the paper plane; thickened black bonds denote chemical substituents projected above the plane, and dashed bonds, beneath the plane (Figure 2). Note that this method requires up/down information for bonds *in addition to* the *x-y* coordinates for atoms.

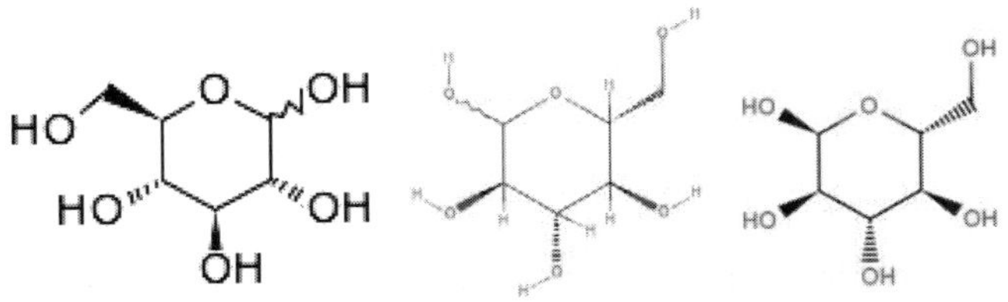

Figure 2. Mills depiction of the same D-glucose from the ligand section of the KEGG database (left), PubChem database (center), and the metabolomics section of the BMRB (right).

1.3. *The Haworth projection*

In glyco- and plant biology, a stereoscopic depiction is far more preferred. In Haworth projection, the sugar ring is placed almost perpendicular to the paper plane, and is viewed from above so that closer atoms and bonds are drawn below the farther components. The orientation usually (but not always) conforms to a clockwise numbering of the IUPAC ring atoms. In Figure 3, for example, CH_2OH group above the ring is C-6. Oxygen is usually placed behind at the right-hand side. The bonds are not necessarily thickened as in Figure 3, and indeed Chemical Abstract Service does not use thickening in its bond description [6].

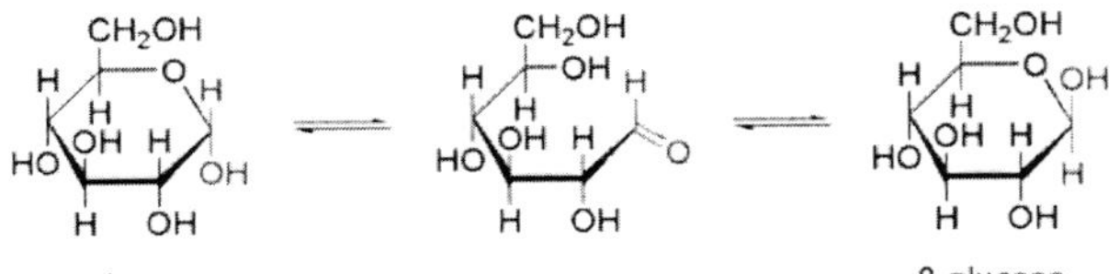

Figure 3. Conversion between *alpha*-D-glucose and *beta*-D-glucose in the Haworth projection. The figure was taken from Wikipedia (Japanese version).

Although the original Haworth projection depicts the ring as planar, it is a highly skewed representation of the original molecular structure. For more precise description, its conformational variant is used to show the boat or chair form. Chemical substituents are called *equatorial* and *axial* when they extend on the plane of the ring, or perpendicular to the ring plane, respectively (Figure 4-a).

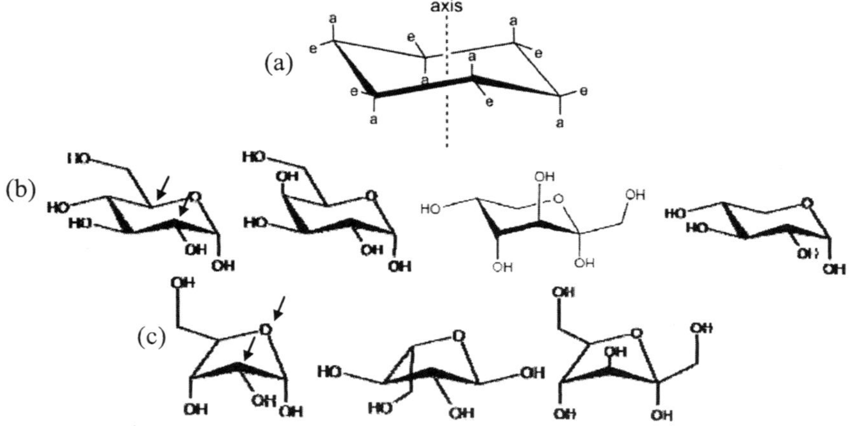

Figure 4. (a) Equatorial and axial directions of the ring, shown with *e* and *a* respectively. (b) Conformations of pyranose sugars. Chair conformation of *alpha*-D-glucose, *alpha*-D-galactose, *alpha*-D-fructose, and *alpha*-D-xylose. In *beta* forms, the hydroxyl group at the lower right corner will become equatorial. We call the two closest carbon atoms (shown with arrows for glucose) as 'concave positions'. (c) Conformations of furanose sugars. From the left, *alpha*-D-ribose, *alpha*-L-arabinose, and *alpha*-D-fructose. For furanose, one carbon and ring oxygen (shown with arrows for ribose) are the concave positions. Note the change between fructopyranose in (b) and fructofuranose in (c), both of which have the same parity. Figures are taken from the website http://www.metabolome.jp/doc/lectures/biochem/sugar/.

Large substituents are thermodynamically more stable when they are equatorial, because of more space in the equatorial positions. By far the most naturally abundant three monosaccharides (glucose, galactose and mannose) have all but one substituent in the equatorial positions in their *beta* forms.

1.4. *Types of monosaccharides*

Most monosaccharides exist as equilibrium mixtures of *alpha*- and *beta*-forms. If we ignore *alpha* and *beta* configuration, four chiral positions for hexoses and three for pentoses exist, theoretically. Each chiral configuration can be represented by a boolean parity corresponding to the Cahn-Ingold-Prelog priority rule (R/S notation) or the D/L nomenclature based on the stereochemistry of glyceraldehydes (visit Wikipedia for these basic rules at http://en.wikipedia.org/wiki/Stereochemistry). Chiral configurations of different monosaccharides are shown in Table 1. The molecular structures for all these monosaccharides in Fischer and Haworth projections are available at our website http://www.metabolome.jp/doc/lectures/biochem/sugar/ in PDF. In the next section, we introduce the procedure to recover these parities from x-y coordinates.

Table 1. Stereo parities of four chiral positions (from C5 to C2) for D-hexose and D-pentose sugars. Corresponding parity for L-sugars can be obtained by flipping all 1 and 2. The parity 1/2 corresponds to right/left orientation of the Fischer projection from the bottom up (see also Figure 1). Nonchiral positions are skipped and the keto group is represented by '_' to distinguish hexoses and pentoses. This parity description from the bottom-up is our original notation.

Sugar Type				
aldohexose (8 patterns)	1111 allose	1121 glucose	1211 gulose	1221 galactose
aldohexose (8 patterns)	1112 altrose	1212 idose	1122 mannose	1222 talose
ketohexose (4 patterns)	111_ psicose	112_ fructose	121_ sorbose	122_ tagatose
aldopentose (4 patterns)	111 ribose	121 xylose	112 arabinose	122 lyxose
ketopentose (2 patterns)	11_ ribulose	12_ xylulose		

2. Algorithms for detecting stereo parities of cyclic monosaccharides

We assume that x-y coordinates and the topology of molecular structure (*i.e.* atoms and their connections) are available. Such structural information includes MOL format and PDB format. SMILES format does not conform to this criterion because it has no coordinates. We also assume that the cyclic parts in the structure are already detected by a standard method in computational chemistry [7].

The detection procedure differs for the Mills depiction, the Haworth projection, and its conformational variant. Since the Mills depiction requires parity information in addition to x-y coordinates, we skip its detection procedure here. The parity information essentially corresponds to the 1/2 codes in Table 1, although a cumbersome parity conversion may be needed depending on the file format of molecular structures. For Mills depiction, therefore, there is no essential need for considering coordinates for sugar detection. We focus here on the remaining two depictions. First, to distinguish the Haworth projection and its conformational variant, distances between all ring atoms that are not directly bonded are computed. In the Haworth projection, there is no single shortest distance because the ring structure is symmetric. An unsymmetrical case is the conformational variant, and we call the closest atom pair 'concave positions' as will be explained in Section 2.3.

2.1. General Strategy

Some hexose sugars may form both pyranose (6-member) and furanose (5-member) rings. We use the notation 'C-x' ($1 \leq x \leq 6$) to refer to the IUPAC carbon positions. Since C-1 carbon (the terminal carbon at the most oxidized side) may be either inside or outside of the ring atoms, our detection algorithm starts from C-5 carbon backward. (Note that C-6 carbon is never included in the ring component for hexoses, and there is no C-6 for pentoses. Starting from C-5 is thus simpler than from C-1 for computational recognition.) To designate ring positions we introduce the following functions.

```
prev(C-x)   := the previous atom in the ring in the IUPAC numbering system;
succ(C-x)   := the next atom in the ring in the IUPAC numbering system;
angle(x, y, p)  := the angle between chemical bond p-x and p-y in radian. Its
```
return value is between $-\pi$ and π.

2.2. Detecting the Haworth projection

We introduce the basic procedure using a computer program pseudo code in `courier` font. Abstract functions are shown in English (times-roman). The actual program code must detect the type of chemical substituents (*e.g.* hydroxyl, keto, amino or other groups) to achieve the final identification of sugars.

Algorithm 2.2
```
int[] parity = new int[4];
q = (Is pentose) ? C-4 : C-5;
for (Carbon position p iterated from q to C-2)
   {
       br = Chemical substituent at p;
       pv = prev(p);
       nx = succ(p);
       parity[p] = (angle(pv, br, p) * angle(pv, nx, p) < 0)
         ? 1 : 2;                                               (1)
       if (Is p on the left of pv)                              (2)
          parity[p] = (parity[p] == 1) ? 2 : 1;
   }
parity[q] = (parity[q] == 1) ? 2 : 1;                           (3)
If (parity[q] == 2)                                             (4)
   for (Carbon position p iterated from q to C-2)
      parity[p] = (parity[p] == 1) ? 2 : 1;
```

Observation: The algorithm 2.2 correctly computes the stereo parities introduced in Table 1 for both hexoses and pentoses.
Proof: Irrespective of the orientation of the Haworth projection, *i.e.*, either clockwise or anticlockwise numbering of ring atoms, the conditional at the line (1) returns 1 when the chemical substituent (*br*) is above the plane of the ring in its upper half of the depiction, and when *br* is below the plane of the ring in its lower half. Therefore if all substituents

are below the plane, the algorithm computes the `parity = { 1, 1, 1, 1 }` in the
`for` loop after correction of the line (2). Only C-5 must be treated differently. Since the
true `prev` position of C-5 is C-6, not the ring oxygen as we defined in the algorithm, the
computed result must be flipped as in the line (3). For example, the algorithm produces
`parity = { 1, 1, 2, 1 }` for D-glucose in Figure 3, which is the same parity as
in Table 1. Finally, the line (4) takes care of all corresponding L-forms.□

Corollary: The algorithm 2.2 is valid even when the ring is in all other, rotated or
flipped orientations.
Proof: It is not hard to see the parity will not change even when the ring is rotated. Let
us consider the flipped situation. Although the line (2) will reverse the parity, since all
chemical substituents become upside down, both changes cancel out to produce the same
parity. Since the conditional at the line (1) returns the same result even when the ring is
flipped over, the final result remains the same.□

2.3   Detecting the conformational variant

The algorithm 2.2 does not work properly for the conformational variant of the Haworth
projection such as in Figure 4-b (please manually simulate the algorithm to understand).
Since chemical substituents are drawn to show equatorial/axial positions, they must be
detected in relation with the plane of the ring, not with the adjacent ring atoms. In the
conformational description, we specify two ring atoms as *concave positions* for hexoses
and pentoses (Figure 4).   These positions are placed in the middle column of the
description and bonds extending from them are always equatorial to the ring plane. To
use such positions we introduce the following functions. Note that a pentose can take
both pyranose and furanose (6- and 5-member rings) forms.

```
concave(C-x)  := return (whether C-x is at the middle position of the ring);
abs(x) := return (x >= 0) ? x : -x;
```

Algorithm 2.3
```
int[] parity = new int[4];
q = If pentose then  C-4 else  C-5;
for  (Carbon position  p  iterated from q to C-2)
    {
      br = Chemical substituent at p;
      pv = prev(p);
      nx = succ(p);
      if (concave(p))
         parity[p]  = (angle(pv, br, p) * angle(pv, nx, p) > 0)
         ? 1 : 2;                                                      (1)
      else {
         cv = adjacent concave atom, either pv or nx;                  (2)
         parity[p]  = (abs(angle(cv, br, p)) < 2π/3) ? 2 : 1  (3)
      }
```

```
    }
```
r = In pyranose, thirdly scanned atom; in furanose, secondly scanned atom.   (4)
```
parity[r] = (parity[r] == 1) ? 2 : 1;
If (parity[q] == 2)
```
   (5)
```
    for  (carbon position  p  iterated from q to C-2)
        parity[p] = (parity[p] == 1) ? 2 : 1;
```

Observation: The algorithm 2.3 correctly computes the stereo parities introduced in Table 1 for hexoses and pentoses.

Proof: The algorithm without the line (4) computes the parity 1 for equatorial, and 2 for axial positions. Note that *beta*-D-glucose (parity 1121) has all its substituents in equatorial position. Similarly, *beta*-D-xylose (parity 121) also has equatorial substituents only (Figure 4). In order to produce the flipped parity, 2 in this case, the line (4) is introduced. We cannot simply use C-3 position instead of r in the line (4) because hexoketose can form both pyranose and furanose (see fructose in Figure 4, for example). The position to be flipped corresponds to the lower left corner of the ring in the configurations of Figure 4, which depends on the size of rings, not on being hexose or pentose. Configurations for all other pentoses and hexoses can be rationalized in relation with the case of glucose, xylose and fructose.□

Corollary: The algorithm 2.3 is valid even when the ring is in all other, rotated or flipped orientations.

Proof: Let us consider *beta*-D-glucose and *beta*-D-xylose again. The algorithm 2.3 recognizes equatorial positions as parity 1 no matter how the structure is rotated or flipped. Because the position to be corrected in line (4) is invariant of rotation or flipping, the final result remains the same.□

Note: The algorithm 2.3 does **not** distinguish D and L series in hexoses. Although the line (5) can deal with L-forms of pentoses, whose ring orientation does not change between naturally occurring D and L series, the L-forms of hexoses change their ring conformation as in Figure 5. Since our algorithm does not consider whether the ring oxygen is at a concave position or not, it can not distinguish D and L series of hexoses. Checking whether the oxygen is concave or convex is necessary (and sufficient) for distinguishing D and L hexoses in Figure 5, but its process becomes highly complex when the ring may be rotated.

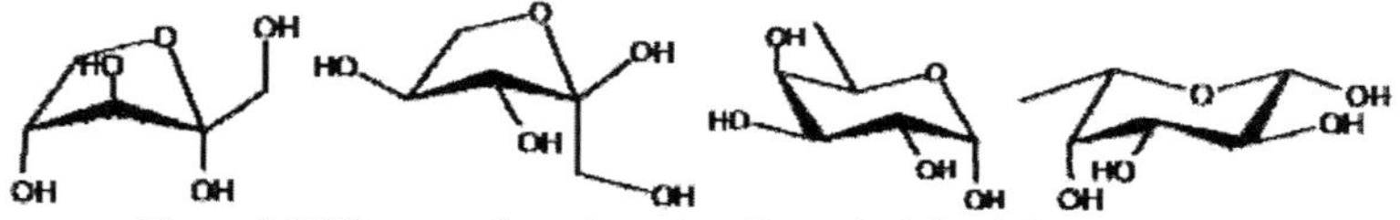

Figure 5. Difference of D and L series. From the left, *alpha*-D-xylulose, *alpha*-L-xylulose (pentose), *alpha*-D-fucose, and *alpha*-L-fucose (hexose).

3. Application to structural data in databases

Flavonoid is a class of secondary metabolites with C6-C3-C6 skeleton derived from phenylpropanoid-acetate pathway [1,8]. Beneficial effect of fruits and vegetables is often attributed to the anti-inflammatory and antioxidant activities of flavonoid, and it has drawn much attention in industry and academia. As shown in the Introduction section, 50 % of registered molecules are glycosides. Since sugars must be attached stepwise by special enzymes called glycosidases, there is a bias for the type of attached monosaccharides in each plant species. Such bias should reflect the evolutionary history of glycosidases, and therefore, of plant families. The algorithms in Section 2 were implemented in Java and were applied to total 6850 natural flavonoid species accessible from our Flavonoid Viewer software at

http://www.metabolome.jp/software/FlavonoidViewer/[†].

This dataset was manually collected from literatures including handbooks of flavonoid and research papers [1,6,8] in collaboration with Kanaya Laboratory in Nara Institute of Science and Technology. The curated structural information is also accessible from the KNApSAcK database of plant secondary metabolites at http://kanaya.aist-nara.ac.jp/KNApSAcK/.

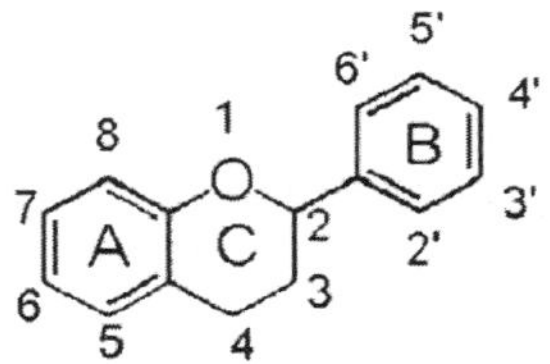

Figure 6. The backbone structure of flavonoid. For isoflavonoid, the phenol ring B is attached at the position 3, not 2. For neoflavonoid, the phenol ring B is attached at the position 4.

In our Flavonoid Viewer software, all molecules are classified into the following 9 categories according to their structure and biosynthetic origin: 1) chalcone and aurone, 2) flavanone, 3) flavone, 4) dihydroflavonol, 5) flavonol, 6) flavan, 7) anthocyanin, 8) isoflavonoid, and 9) neoflavonoid. For each of these categories, we identified monosaccharide types that are directly attached to their backbone structure.

3.1. Pattern of glycosylation

Major monosaccharides directly attached to the flavonoid backbone are listed in Table 2. From the Table, we can immediately tell the uniqueness of anthocyanins. Most common are their 3-O-glucosides, 3-O-galactosides, and 5-O-glucosides. The rest of the

---

[†] The software is freely accessible, but it is unpublished and the structural data are currently not downloadable in a bulk. Users can search and view, however, all structural data.

categories are much less glycosylated, but still show category-dependent characters. In flavones, 6-C-glucosides and 8-C-glucosides are observed. These molecular species (e.g. isovitexin C-glucosides) are mainly identified in caryophyllaceae (carnation family), but also occur in many higher plants such as fabaceae (bean family), gentianaceae (gentian family), or passifloraceae (passion flower family). The C-glucosylation should occur after oxidation of flavanones because the C-glucosides are not common in flavanones. Another character is the abundance of 7-O-arabinosides. These observations coincide with a previous report [14]. The characteristic of flavonols is 3-O- and 7-O-rhamnosylation. The abundance of 3-glycosides in flavans and anthocyanins in Table 2 suggests either these modifications occur prior to reduction of flavonols into flavans, or the substrate specificity of 3-O-rhamnosidase is not tight (the latter is the current consensus of experts). Such observations have been partially described previously [10], but this is the first solid statistical result obtained from around 7,000 natural molecular structures identified to date. In summary, our algorithms produced statistics only from molecular structures, and the obtained results coincided with previous observations by experts.

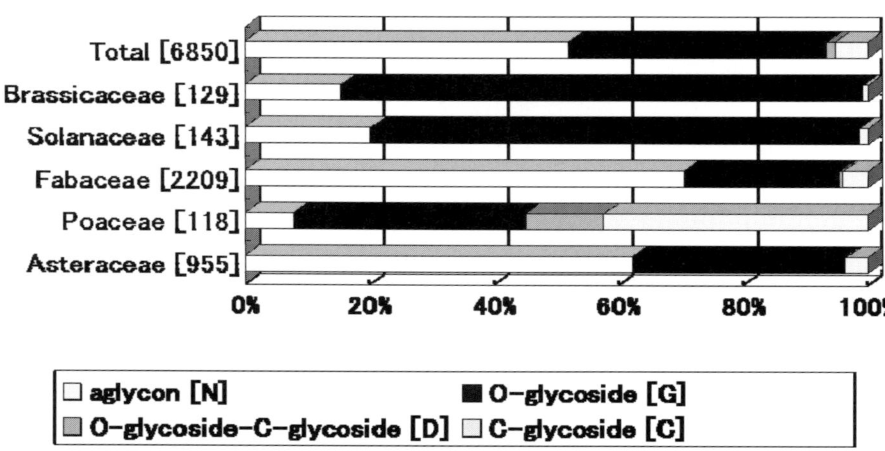

Figure 7. The glycosidation patterns for major plant families. Aglycon means backbone structure without sugars, and O-glycoside-C-glycoside means structure with both O- and C-glycosides. See main text for plant family names.

3.2. Pattern of Glycosylation in Plant Families

Using the species-molecule relationships obtained from the KNApSAcK database for plant secondary metabolites [13], we can obtain the distribution of flavonoid categories in plant families. As shown in Figure 7, we could confirm the following observations: 1) C-glycosides are abundant in poaceae (rice family), whereas O-glycosides are major in other representative plant families such as asteraceae (chrysanthemum family), solanaceae (eggplant family), brassicaceae (cabbage family), and fabaceae (bean family); 2) More than 95 % of isoflavonoids occur in fabaceae (statistical data not shown).

Table 2. Summary of the types of monosaccharides directly attached to the backbone in Figure 6. Components of less than 2% are ignored. The value in parentheses is the number of molecules currently registered in the software. All structures are accessible on the Internet. Abbreviations: Glc ... D-glucose; Gal ... D-galactose; Rha ... D-rhamnose.

Flavonoid Category	Monosaccharide Type at Each Position
chalcone and aurone (690)	Position 5: Glc 2%
	Position 7: Glc 6%; Gal 3%
flavanone (698)	Position 4': Gal 2%
	Position 5: Glc 2%
	Position 7: Glc 8%; Gal 4%
flavone (1475)	Position 4': Gal 4%
	Position 6: Glc 9%; Gal 3%
	Position 7: Glc 14%; Gal 4%; L-arabinose 2%
	Position 8: Glc 6%; Gal 3%
dihydroflavonol (272)	Position 3: Rha 6%; Glc 4%
	Position 7: Glc 4%
flavonol (1942)	Position 4': Glc 2%
	Position 3: Glc 21%; Gal 11%; Rha 10%
	Position 7: Glc 10%; Rha 3%; Gal 2%
flavan (289)	Position 3: Glc 2%; Rha 2%
	Position 5: Glc 5%
	Position 7: Glc 5%
anthocyanin (486)	Position 3': Gal 2%; Glc 2%
	Position 5': Gal 3%
	Position 3: Glu 64%; Gal 20%; Rha 2%
	Position 5: Glu 33%; Gal 8%
	Position 7: Gal 5%; Glc 4%
isoflavonoid (916)	Position 7: Glc 9%
	Position 8: Glc 2%
neoflavonoid (82)	Position 5: Glc 7%; Gal 3%

4.   Conclusion

In the past decade we have seen an almost explosive expansion of biological databases. We demonstrated here that characteristics of plant families can be obtained from unbiased statistic of species-molecule information computationally obtained from well-curated structural databases. Database accuracy is of crucial importance. In this perspective, our computational procedure contributes not only to sort out polysaccharide structures but also to detect input errors (e.g. molecular names or structures) in glycomics databases. We propose that all polysaccharide information in databases be checked computationally to improve its data quality and to transfer its value to other databases with pathway and other metabolic information. Our future work is therefore to extend the structural recognition system to major acylation patterns in secondary metabolites such as acetyl, caffeic, p-coumaric, ferulic, gallic, malic, and malonic acids.

Acknowledgments

We thank Kazuhiro Suwa for the implementation of Flavonoid Viewer, and Yukiko Nakanishi and Yukiko Fujiwara for curating structural data of flavonoids. We also thank Yoko Shinbo and Prof. Shigehiko Kanaya for providing us the structure data of flavonoids and species-molecule relationships from the KNApSAcK database. This work is a part of joint research with Kanaya Laboratory, and is supported by Grant-in-Aid for Scientific Research on Priority Areas "Systems Genomics" from the Ministry of Education, Culture, Sports, Science and Technology of Japan.

References

[1]   Andersen, O.M., Markham, K.R. (eds.) Flavonoids: chemistry, biochemistry and applications. CRC Press, 2006.

[2]   Apweiler, R. On the frequency of protein glycosylation, as deduced from analysis of the SWISS-PROT database, *Biochim. Biophys. Acta* 1473:4–8, 1999

[3]   Banin, E., Neuberger, Y., Altshuler, Y., Halevi, A., Inbar, O., Dotan, N. and Avinoam, D. A novel linear code nomenclature for complex carbohydrates, *Trends Glycosci. Glycotechnol.* 14(77):127-137, 2002.

[4]   BMR Data Bank: http://www.bmrb.wisc.edu/metabolomics/

[5]   Bohne-Lang, A., Lang, E., Forster, T., and von der Lieth, C. W. LINUCS: linear notation for unique description of carbohydrate sequences, *Carbohydr. Res.* 336: 1-11, 2001.

[6]   Chemical Abstract Service: http://www.cas.org/

[7]   Gasteiger, J. (ed.) Handbook of Chemoinformatics: From Data to Knowledge Vol. 1, John Wiley & Sons Inc. 2003.

[8]   Harborne, J. B. and Baxter, H. (eds.) The handbook of natural flavonoids. John Wiley & Sons, 1999.

[9]   IUPAC-IUBMB Nomenclature of Carbohydrates 2-Carb-5: http://www.chem.qmul.ac.uk/iupac/2carb/05.html

[10] Iwashina, T. The structure and distribution of the flavonoids in plants. *J. Plant Res.* 113, 287-299, 2000.

[11] KEGG Database: http://www.genome.ad.jp/kegg/

[12] KEGG Glycan database: http://www.genome.jp/kegg/glycan/

[13] KNApSAcK database http://kanaya.aist-nara.ac.jp/KNApSAcK/

[14] Mastenbroek, O., Prentice, H.C., Kamps-Heinsbroek, R., van Brederode, J., Niemann, G. J., van Nigtevecht, G. Geographic trends in flavone-glycosylation genes and seed morphology in EuropeanSilene pratensis (Caryophyllaceae) *Plant Systematics and Evolution* 141(3-4), 257-271, 1983.

[15] PubChem Database: http://pubchem.ncbi.nlm.nih.gov/

[16] Tokimatsu, T. and Arita, M. unpublished results. Review in Japanese is available as Tokimatsu, T. and Arita, M. Viewing flavonoid through metabolic maps. *Saibou Kougaku*, 25(12) 1388-1393, 2006.

MINING SUPER-SECONDARY STRUCTURE MOTIFS FROM 3D PROTEIN STRUCTURES: A SEQUENCE ORDER INDEPENDENT APPROACH

ZEYAR AUNG[1]

azeyar@i2r.a-star.edu.sg

JINYAN LI[2]

jyli@ntu.edu.sg

[1] *Institute for Infocomm Research, 21 Heng Mui Keng Terrace, Singapore 119613*

[2] *School of Computer Engineering, Nanyang Technological University, Nanyang Avenue, Singapore 639798*

Super-Secondary structure elements (super-SSEs) are the structurally conserved ensembles of secondary structure elements (SSEs) within a protein. They are of great biological interest. In this work, we present a method to formally represent and mine the sequence order independent super-SSE motifs that occur repeatedly in large data sets of protein structures. We represent a protein structure as a graph, and mine the common cliques from a set of protein graphs in order to find the motifs. We mine two categories of super-SSE motifs: the generic motifs that occur frequently across the entire database of protein structures, and the fold-preferential motifs that are concentrated in particular protein fold types. From the experimental data set of 600 proteins belonging to 15 large SCOP Folds, we have discovered 21 generic motifs and 75 fold-preferential motifs that are both statistically significant and biologically relevant. A number of the discovered motifs (both generic and fold-preferential) resemble the well-known super-SSE motifs in the literature such as beta hairpins, Greek keys, zinc fingers, etc. Some of the discovered motifs are of novel shapes that have not been documented yet. Our method is time-efficient where it can discover all the motifs across the 600 proteins in less than 14 minutes on a standalone PC. The discovered motifs are reported in our project webpage:
http://www1.i2r.a-star.edu.sg/~azeyar/SuperSSE/

Keywords: 3D Protein Structure, Super-secondary Structure, Structural Motifs Mining.

1. Introduction

Proteins are the workhorses in the cells of living organisms. A protein is made up of a sequence of amino acid (AA) residues which folds into a particular 3-dimensional (3D) structure by the various forces of nature. A 3D protein structure consists of frequent and structurally conserved elements called secondary structure elements (SSEs). Alpha helices and beta strands are the two common types of SSEs. There are in turn some ensembles of SSEs that are frequent and structurally conserved. They usually serve as the structural and/or functional units within a protein, and are called super-secondary structure elements (super-SSEs) [4].

Biologists are very interested in super-SSEs because they are usually associated with basic structural configurations and/or basic biological functions of the proteins.

Some of the well-known super-SSE types are helix-loop-helix, beta ribbon, beta hairpin, beta-alpha-beta, zinc finger, EF hand, Greek key, etc. Researchers have studied super-SSEs extensively for more than three decades [12, 21, 23–26].

A super-SSE motif is a particular type of structurally similar super-SSEs that occur frequently across a given set of protein structures. In this paper, we propose a method to (1) formally represent the sequence order independent (sequentially disconnected) super-SSEs with respect to their structural conformations, and (2) mine the motifs of those super-SSEs in a given set (either the entire database or a particular fold type) of protein structures.

Conventionally, a super-SSE is defined as a set of sequentially connected (i.e. sequence order preserved) SSEs that are neighbored to each other in 3D space. However, there exists a number of biologically significant structural motifs being composed of SSEs that are spatially proximate yet sequentially not connected [5, 9, 25]. Such a motif can be termed a *sequence order independent* motif.

In this work, we generalize the definition of a super-SSE by relaxing the sequence order constraint with a view to covering the sequence order independent motifs. For example, while the conventional definition covers only the sequence order preserved motif A–B–C as shown in Fig. 1(a), our definition can also deal with the sequence order independent motif A'–B'–C' as shown in Fig. 1(b).

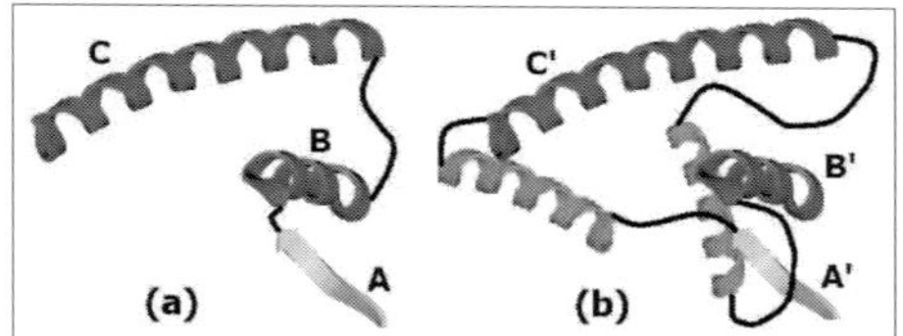

Fig. 1. (a) A conventional (sequence order preserved) beta-alpha-alpha super-SSE. (b) A sequence order independent super-SSE with the same spatial configuration.

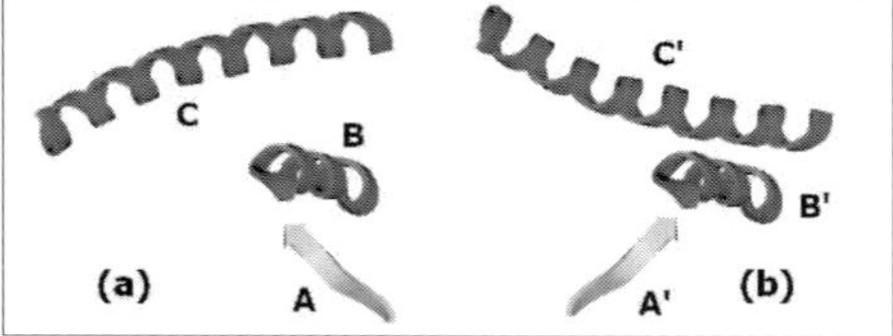

Fig. 2. Two beta-alpha-alpha super-SSEs with different structural configurations.

In our proposed method, we represent a protein structure as a labeled graph with each node being an SSE, and each edge being the relationship between two close enough SSEs. A clique (a fully connected sub-graph) within a graph corresponds to a super-SSE. We develop an algorithm to mine the frequent clique types (super-SSE motifs) in a given protein structure data set. From a experimental data set of 600 proteins, we can discover a number of generic and fold-preferential motifs that are both statistically significant and biologically relevant within a short time.

2. Motivations

2.1. *Need for a Formal Representation Scheme*

Traditionally, super-SSEs (both sequence order preserved and sequence order independent) are described less formally with the names such as helix-loop-helix, alpha-beta-alpha, etc. Using this verbal description, we have only a very limited ability to identify and quantify the super-SSEs systematically. For example, we may be able to distinguish a beta-alpha-alpha from an alpha-beta-beta. But we may not be able to differentiate between two beta-alpha-alpha super-SSEs having different structural configurations as shown in Fig. 2. The ability to distinguish or classify such kinds of SSEs into different types is highly desirable for the biologists, since it will enable them to study super-SSEs in a more subtle manner [12].

Some methods such as [12, 24] try to identify the different types of super-SSEs by characterizing the loops between the constituent SSEs of a super-SSE. But, this approach is also limited because it requires the sequence order constraint, and its applicability is confined to the super-SSEs with only two elements.

Thus, there is a need for a formal representation scheme which enables the identification and quantitative manipulation (comparison, clustering, etc.) of super-SSEs in a more general manner (i.e. applicable to all kinds of super-SSEs regardless of their sequence order and the number of SSEs they contain). In this work, we try to address this formalization issue by representing proteins and super-SSEs as labeled graphs and labeled cliques respectively.

2.2. *Need for a Large-Scale Motif Mining Method*

Structural motif mining is an active area of research in structural bioinformatics. Different methods use different description of structural motifs, and try to mine the frequent motifs from a set of protein structures. Trilogy [3] explores the sequence–structure motifs made up of AA residue triplets; SPratt2 [17] mines the conserved residues within a fixed-size bounding sphere; MotifMiner [6] mines the frequent atom-sets; and Huan *et al.* [15, 16] mines the frequent sub-graphs/cliques of AA residues, etc.

In this study, we will focus on the discovery of structural motifs in terms of the super-SSEs. A number of methods, such as Koch *et al.* [19], MASS [9], PROTEP [1], and Szustakowski *et al.* [25], have been proposed to detect both sequence order preserved and sequence order independent super-SSE motifs. All of these methods adopt the *comparison-based* motif discovery approach [11] in which each method employs one of the many multiple structural alignment algorithms to generate the motifs. Unfortunately, such a comparison-based approach is only suitable for the discovery of motifs from small data sets with just tens of protein structures. In terms of its scalability, it is not suited for motif discovery from larger data sets with hundreds or thousands of proteins for the following reasons.

- Usually, a motif does not occur in all proteins in the data set, but only in a subset of it. Since we do not know *a priori* the motifs nor the subsets of proteins in which these motifs occur, we need to explore all the possible combinations of proteins in the data set. In order to retrieve the complete set of motifs from a given set of N proteins, a naive approach will take an exponential time, whist an intelligent approach, such as the one described in Koch *et al.* [19], will still take an $O(N^3)$ time.

- If a greedy strategy is adopted to reduce the time cost, some pivot proteins can be selected to serve as seeds for multiple alignments. But, this cannot always guarantee a complete answer in the event where a motif does not occur in any of the selected pivots.
- Although a single run of an expensive comparison-based motif discovery algorithm (which may take several days to several weeks) may still be affordable, one will need multiple runs of the algorithm with different sets of parameters in order to secure the desired results. Such multiple runs are prohibitively expensive to be carried out in reality.

With a view to overcoming the abovementioned problems, we adopt the *pattern-mining* approach — also known as the *pattern-driven* approach [11] — for the large-scale discovery of super-SSE motifs. The pattern-mining strategy has been used for discovering sequence, structure, and sequence–structure motifs of various kinds [3, 6, 15–17]. However, to our best knowledge, it has not been used for the discovery of super-SSE motifs before.

Since we represent a protein structure as a graph, we need to apply pattern mining algorithms for graphs so as to discover our desired super-SSE motifs from the graphs. This has been technically infeasible until the recent emergence of the algorithms for the large-scale mining of graph databases for sub-graphs [28], quasi-cliques [22], and cliques [27].

In this work, we utilize one of these latest technologies, namely CLAN [27], to mine the frequent cliques representing the super-SSE motifs. CLAN is known to be a complete clique mining algorithm where it enumerates all the frequent closed cliques from a given database of graphs. It is also an efficient tool that can manage large graph databases with fast response times.

Recently, Huan *et al.* [15, 16] has used graph representation and mining to find the motifs of AA residue nodes. However, it should be noted that their objective is substantially different from ours in which they try to mine the small residue-based packing motifs rather than the relatively large super-SSE motifs as in our case.

3. Methods

3.1. *Formal Representation Of Super-SSE Motifs*

In this section we will describe how we formalize the representation of a protein and that of a super-SSE motif.

3.1.1. *SSE as a Vector*

We use the STRIDE algorithm [13] to identify the SSEs in protein structures. Since SSEs are relatively straight in structure, we can approximate each SSE with a vector (line segment) in 3D space [9, 21]. Fig. 3(b) shows the vector representation of SSEs.

3.1.2. *Protein Structure as a Graph*

We present a protein structure as a graph with its nodes being the SSE vectors, and edges being the relationships between these SSE vectors. Graph representation of protein structures has also been used previously in a number of protein structure comparison and analysis methods [1, 19, 21].

For a protein with n number of SSEs, we have a graph of n nodes. A pair of nodes in the graph is connected by an edge if the *distance of closest approach* [31] between the corresponding SSE vector pair is less than the distance threshold dt. The constituent SSEs in a super-SSE must be close enough to each other, i.e. less than dt, in order to act effectively as a structural/functional unit. Since we do not put an edge between any pair of nodes whose SSE vectors are farther than dt, those two SSEs can never become parts of a single super-SSE. We use $dt = 16$Å as the default value. The graph representation of a protein structure is depicted in Fig. 4.

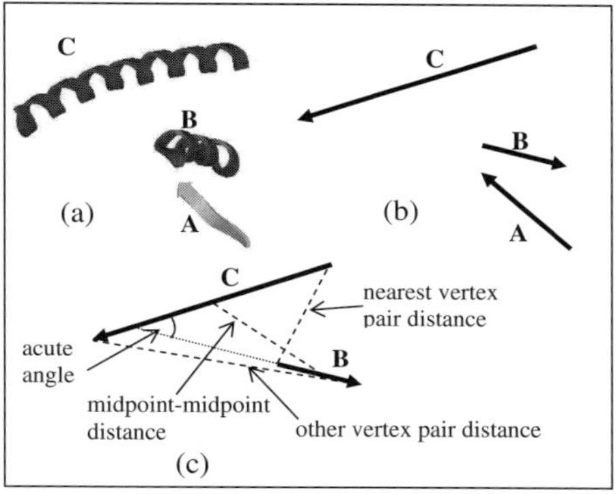

Fig. 3. (a) Original SSEs (b) Vector representation of SSEs and (c) Various types of relationships between SSEs.

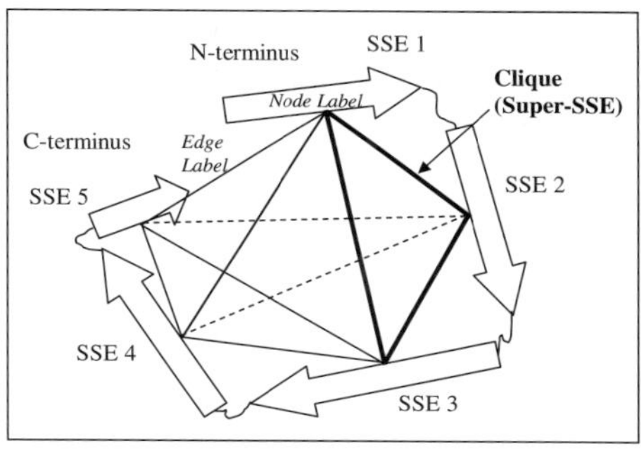

Fig. 4. Graph representation of a protein structure with 5 SSEs. A dotted line denotes a non-existing edge between two node because their SSEs are farther than the distance threshold dt.

Labels are assigned to all nodes and edges. Each node label corresponds to the attributes of the SSE it represents. We use two attributes: (1) type (alpha-helix or beta-strand) and (2) length (in terms of the number of AAs) of the SSE for a node label. Each edge label corresponds to the attributes of the relationship of the two SSEs it connects. We use four attributes: (1) acute angle, (2) nearest vertex-pair distance, (3) other vertex-pair distance, and (4) midpoint-midpoint distance between the two SSE vectors for an edge label. Fig. 3(c) demonstrates the four edge label attributes. Our graph representation scheme is sequence order independent in that the node and the edge labels do not carry any information regarding sequence positions or sequential connectivity of the SSEs.

For each label (either for node or edge), we quantize each attribute, concatenate the binary values for all attributes, and convert the concatenated bit string into a single integer value. The number of bins for each attribute is empirically determined.

3.1.3. *Super-SSE Motif as a Clique*

In a graph for a protein structure as described above, each *clique* (a sub-graph where every node is connected by an edge with every other node) can be viewed as a super-SSE. According to our definition, every constituent SSE within a super-SSE must be close enough (i.e. connected by an edge) to every other constituent SSE. Thus, any other kind of non-clique induced sub-graph does not qualify as a super-SSE.

Two given cliques (representing two super-SSEs) can be considered as structurally similar and thus belonging to the same type if they are isomorphic, i.e. all of their corresponding node and edge labels are matched. (Partially-matched cliques are not guaranteed to similar to each other despite their matching portions.) If the instances of a particular super-SSE type occurs *frequently* in a given set of protein, this super-SSE type can be defined as a *motif*.

3.2. *Mining Super-SSE Motifs*

A frequent clique is a clique that occurs in at least *st* graphs in a given set of protein structure graphs, where *st* is a user-defined support threshold. A frequent clique corresponds to a super-SSE motif.

We find the frequent cliques from the given set of graphs using a general-purpose frequent clique mining algorithm called CLAN [27]. CLAN reports the frequent cliques only in terms of their node labels. (Hereafter, we will name such a clique as a *node-frequent* clique.) In other words, the set of node-frequent cliques reported by CLAN is a superset of the set of actual frequent cliques with both their node and edge labels taken into account.

Thus, we have to test whether a clique reported by CLAN is actually frequent or not. Since CLAN reports only the node labels of the node-frequent cliques and their respective support values, we have to find the actual instances of these node-frequent cliques in all the protein graphs in the data set. We use the VF2 [7] sub-graph isomorphism algorithm to find these instances.

After finding all the instances for a node-frequent clique in all protein graphs, we find the frequent instance(s) (both in terms of their node and edge labels) that occur in at least *st* protein graphs in the given data set, and report them as the desired super-SSE motifs. (Note that, for one node-frequent clique, there may be more than one distinct frequent clique because of the different edge labels. On the other hand, for some node-frequent cliques, there may be no actual frequent clique at all. It was observed that the number of the actual frequent cliques is only about 10% of the original node-frequent cliques.)

We try to find two categories of super-SSE motifs: (1) the motifs that occur frequently across the entire database — termed the *generic motifs*, and (2) the motifs that occur concentratively in particular protein fold types (SCOP Folds in our case) — termed the *fold-preferential motifs*.

3.2.1. *Generic Motifs*

First, we find the generic motifs each of which occurs in at least st_g proteins across the whole given database of protein structures, where st_g is a user-defined support threshold. After we have discovered the generic motifs by the procedure described above, we need to assess their statistical significance. For that, we calculate the estimated p-values of them using the model described by He and Singh [14].

According to this model, we can represent a generic motif ω as a feature vector of the occurrences of the basic elements it contains.

$$\omega = \{y_1, y_2, \ldots, y_t\} \tag{1}$$

where t is the number of unique basic elements in the database, and $y_i\,(1 \leq i \leq t)$ is the number of occurrences of the i-th basic element in the motif ω.

Here, we treat each distinct *combined label*, which is a concatenated string of the label of an edge plus the labels of nodes connected by the edge, as our basic element. We can calculate the probability of ω occurring at random in a protein graph in the database as:

$$\hat{P}(\omega) = \prod_{i=1}^{t} P(Y_i \geq y_i) \tag{2}$$

where $P(Y_i \geq y_i)$ is the probability that the i-th basic element (combined label) occurs at least y_i times in a random vector. This is calculated based on the background distribution of the basic elements in the database. Finally, the p-value of the generic motif ω (termed *generic p-value*) is calculated as:

$$PV_g(\omega) = \sum_{\mu=T}^{N} bino(\mu, N, \hat{P}(\omega)) \tag{3}$$

where N is the number of graphs (proteins) in the database; T is the support, i.e, the number of proteins in which the motif ω occurs $(T \geq st_g)$; and $bino(.,.,.)$ is the binomial distribution function. If the generic p-value is less than or equal to 0.05, the motif is considered statistically significant.

3.2.2. *Fold-preferential Motifs*

Second, we mine the fold-preferential motifs that occur more frequently in a certain protein fold type rather than in the other protein fold types. In particular, we find the motifs that are concentrated in certain SCOP Folds. (SCOP [30] is a protein structure classification system. A Fold in SCOP consists of a set of proteins that are generally similar to each other in terms of their 3D structures.) We define a particular motif as fold-preferential only if the motif occurs in at least twice the number of proteins in its most frequent SCOP Fold than in its second-most frequent SCOP Fold.

We find the fold-preferential motifs each of which occurs in at least st_f proteins in its most frequent SCOP Fold, where st_f is a user-defined support threshold. Then, we calculate the statistical significance of the fold-preferential motifs in terms of another type of p-value named *fold-preferential p-value.*

We can calculate the fold-preferential p-value of a particular motif ω to occur by chance in a particular SCOP Fold by using a hypergeometric distribution [15]:

$$PV_f(\omega) = 1 - \sum_{i=0}^{K-1} \frac{\binom{F}{i}\binom{N-F}{T-i}}{\binom{N}{T}} \tag{4}$$

where N is the number of proteins in the entire database; T is the total number of proteins in which the motif occurs in the entire database; (For each motif for a SCOP Fold, we also have to enumerate its other instances outside its own Fold in the rest of the database by using the VF2 algorithm again.) F is the size of the SCOP Fold in which the motif most frequently occurs; and K is the number of proteins in which the motif occurs in this Fold ($K = T \cap F$). Again, if the fold-preferential p-value is less than or equal to 0.05, the motif is regarded as statistically significant.

4. Results and Discussions

We use the same database of 600 proteins as previously used in [2]. The list of the 600 proteins is given in the project webpage. This is a subset of the SCOP database [30] with less than 40% sequence homology. The PDB-style co-ordinates for these proteins are obtained from the ASTRAL database [29].

The database of 600 proteins is composed of 15 large SCOP Folds each having 40 member proteins. (If a Fold contains more than 40 members, we randomly select 40 from it.) The SCOP designations for these 15 Folds and their descriptions are given in Table 1.

First, we mine the generic super-SSE motifs that occur frequently across the whole database of 600 proteins with the support threshold of $st_g = 3\%$, and assign the generic p-values to the motifs. Then, we find the fold-preferential super-SSE motifs for each of the 15 SCOP Folds with the support threshold $st_f = 10\%$, and assign both the fold-preferential p-values and the generic p-values to the motifs.

We conducted our experiments on a single PC with Pentium D 3.2GHz processor and 2GB main memory running Windows XP. The time statistics show that the proposed method is efficient. The total running time using the default parameters ($dt = 16\text{Å}$, $st_g = 3\%$, and $st_f = 10\%$) is only 805 sec (13 min 25 sec) in which 178 sec is for constructing the protein structure graphs, 274 sec is for mining of the generic motifs, and 353 sec for mining the fold-preferential motifs.

The effects of varying the three important parameters dt, st_g and st_f are discussed in the project webpage.

4.1. *Generic Motifs*

We have discovered a total of 22 generic motifs among which 21 are statistically significant in terms of their generic p-values. All of these 21 generic motifs are 3-SSE

Table 1. Number of significant fold-preferential motifs discovered in the SCOP Folds ($dt = 16$Å, $st_f = 10\%$).

SCOP Fold	Description	#3-SSE motifs	#4-SSE motifs	Total
a.4	DNA/RNA-binding 3-helical bundle	0	0	0
a.39	EF Hand-like	0	0	0
a.118	alpha-alpha superhelix	0	0	0
b.1	Immunoglobulin-like beta-sandwich	1	0	1
b.40	OB-fold	0	0	0
c.1	TIM beta/alpha-barrel	8	0	8
c.2	NAD(P)-binding Rossmann-fold domains	20	3	23
c.3	FAD/NAD(P)-binding domain	13	4	17
c.23	Flavodoxin-like	0	0	0
c.37	P-loop containing nucleoside triphosphate hydrolases	1	0	1
c.47	Thioredoxin fold	3	0	3
c.55	Ribonuclease H-like motif	0	0	0
c.69	alpha/beta-Hydrolases	20	1	21
d.15	beta-Grasp (ubiquitin-like)	1	0	1
d.58	Ferredoxin-like	0	0	0
Total		67	8	75

motifs. (There are a vast number of 2-SSE motifs. In this work, we simply ignore them because they are considered less significant. On the other hand, we have not detected any frequent motif with the size larger than 3 SSEs.) We rank the motifs by their generic p-values. The distribution of the motifs' generic p-values is shown in Fig. 5.

The highest-ranked generic motif has the lowest p-value of 5.75×10^{-22}. Its random probability $\hat{P}(\omega)$ is 0.0272, and it occurs in 67 proteins across 7 distinct SCOP Folds. It resembles a version of a well-known conventional super-SSE motif called three-stranded beta hairpin [8, 10] with all beta strands approximately parallel to each other as shown in Fig. 7. (Higher-resolution images for Fig. 7–10 can be viewed in the project webpage.)

We have also discovered a number of other biologically relevant motifs that look

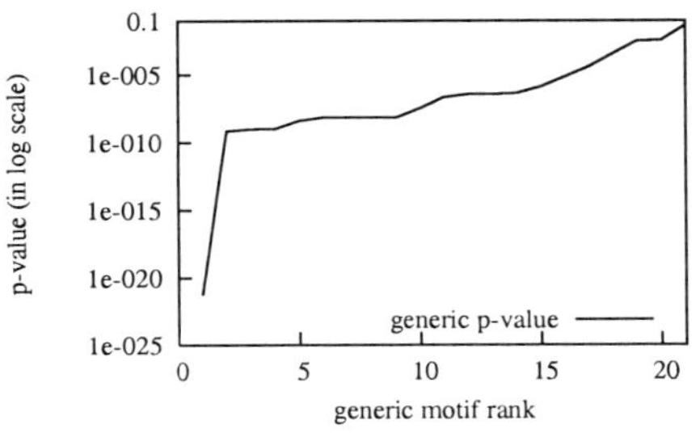

Fig. 5. P-values of generic motifs ($dt = 16$Å, $st_g = 3\%$).

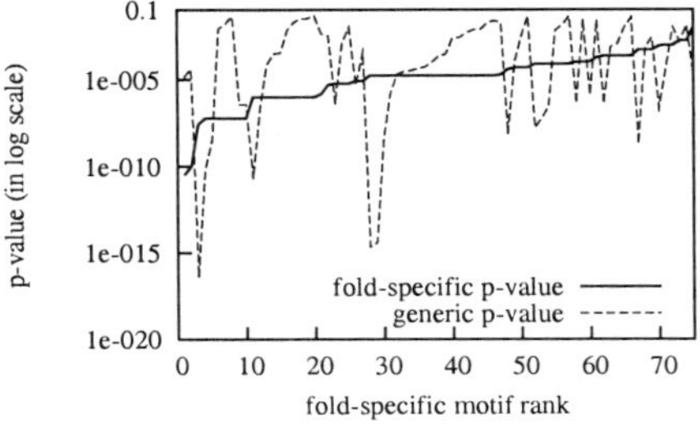

Fig. 6. P-values of fold-preferential motifs ($dt = 16$Å, $st_f = 10\%$).

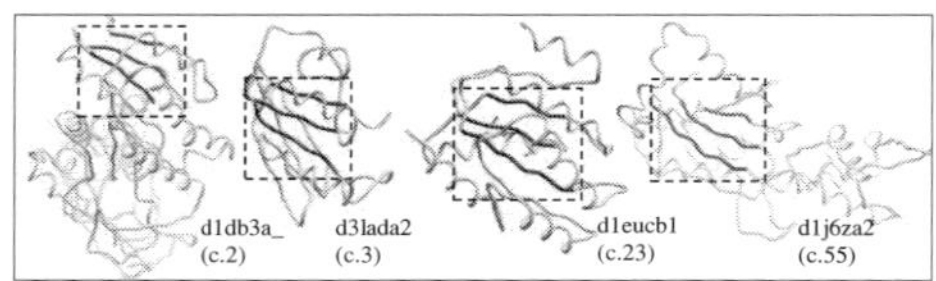

Fig. 7. Some instances of the rank #1 generic motif: a 3-SSE motif resembling a three-stranded beta hairpin with all parallel beta strands.

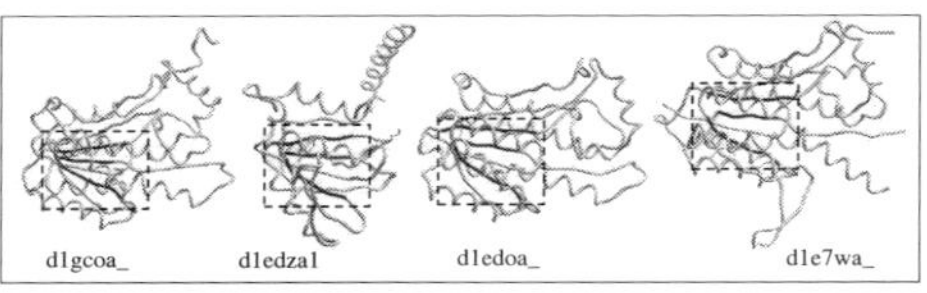

Fig. 8. Some instances of the rank #1 fold-preferential motif in SCOP Fold c.2: a 3-SSE motif resembling a three-stranded beta hairpin with two parallel and one angled beta strands.

like the well-known conventional super-SSE motifs such as different versions of beta hairpins, beta-alpha-beta, zinc fingers, etc. The complete list of 21 generic motifs and their occurrences is reported in the project webpage.

4.2. *Fold-preferential Motifs*

We have found a total of 110 fold-preferential motifs among which 75 are statistically significant in terms of both their fold-preferential and generic p-values. Among these 75 significant motifs, 67 are the 3-SSE and 8 are the 4-SSE motifs. 9 of the fold-preferential motifs overlap with the generic motifs.

The motifs are found in 8 out of the 15 SCOP Folds investigated. The number of motifs found for each Fold is given in Table 1. We rank the motifs by their fold-preferential p-values. The distributions of the p-values of both kinds for those 75 motifs are shown in Fig. 6.

The highest-ranked motif has the lowest fold-preferential p-value of 3.22×10^{-11}, and the genetic p-value of 1.40×10^{-5}. It is preferential to SCOP Fold c.2. It occurs in 10 proteins in c.2, but only in 2 proteins in the rest of the database. It is also similar to a version of the three-stranded beta hairpin motif [8, 10] with two parallel and one angled beta strands as shown in Fig. 8.

We have found a 4-SSE motif as our third-ranked motif. It is a beta-beta-beta-alpha motif (Fig. 9) which resembles the sequence order preserved version described in [18]. It has the fold-preferential p-value of 2.56×10^{-8}, and the genetic p-value of 3.80×10^{-17}. It is preferential to SCOP Fold c.3. It exists in 7 proteins in c.3, but only in 1 protein in the rest of the database.

We have also discovered a number of other biologically relevant motifs as our fold-preferential motifs. We report the full list of the 75 fold-preferential motifs in the project webpage.

It is observed that we have achieved our objective of formalization and specification of the super-SSE motifs as discussed in Section 2.1. Different versions of the motifs with the same verbal description can be classified based on their structural configurations. For example, we are able to distinguish the two different versions of the 3-SSE motifs resembling the three-stranded beta hairpin [8, 10] as shown in Figures 7 and 8. It has been previously observed that super-SSEs or SSE packings

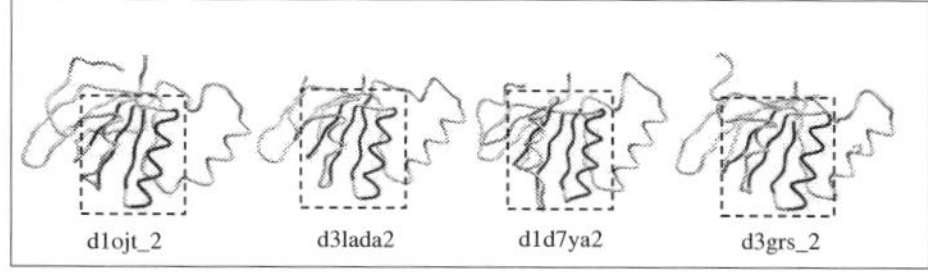

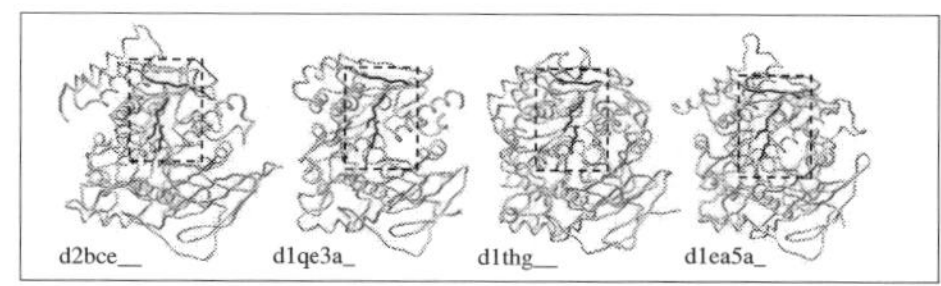

Fig. 9. Some instances of the rank #3 fold-preferential motif in SCOP Fold `c.3`: a 4-SSE motif resembling a beta-beta-beta-alpha motif.

Fig. 10. Instances of a 3-SSE T-shape fold-preferential motif in SCOP Fold `c.69` (fold-preferential rank #44).

with the same SSE components but with the different configurations correspond to different biological functions [12, 20]. As such, it can be conjectured that those different versions of motifs may have different functions. However, an in-depth biological analysis will be required to verify this.

In addition, we have also discovered some new types of super-SSE motifs (both generic and fold-preferential) whose shapes have not been documented in the literature yet. For example, we have discovered a 3-SSE T-shape motif preferential to SCOP Fold `c.69` as shown in Fig. 10. Biologists can further investigate their detailed structural and functional properties, and possibly explore their potential usability in biomedical applications such as drug target finding.

5. Conclusion

In this paper, we have proposed a method to formalize the representation of sequence order independent super-SSEs, and mine the frequent super-SSE motifs from a large data set of protein structures. We have shown that our method is both effective and efficient. It can discover the generic and fold-preferential motifs that are statistically significant and biologically interesting within a short time. Biologists can further explore our discovered motifs to find out the potential usability of them in biomedical applications.

References

[1] Artymiuk, P. J., Spriggs, R. V., and Willett P., Graph theoretic methods for the analysis of structural relationships in biological macromolecules, *J. Am. Soc. Info. Sci. Tech.*, 56:518–528, 2005.

[2] Aung, Z. and Tan, K. L., Automatic 3D protein structure classification without structural alignment, *J. Comp. Biol.*, 12:1221–1241, 2005.

[3] Bradley, P., Kim, P. S., and Berger B., TRILOGY: discovery of sequence–structure patterns across diverse proteins, *Proc. Natl Acad. Sci., USA*, 99:8500–8505, 2002.

[4] Branden, C. and Tooze, J., *Introduction to Protein Structure*, Garland Publishing, 2nd edition, 1999.

[5] Chothia, C., Levitt, M., and Richardson, D., Structure of proteins: packing of alpha-helices and pleated sheets, *Proc. Natl Acad. Sci., USA*, 74:4130–4134, 1977.

[6] Coatney, M. and Parthasarathy, S., MotifMiner: efficient discovery of common substructures in biochemical molecules, *Knowl. & Inform. Sys.*, 7:202–223, 2005.

[7] Cordella, L. P., Foggia, P., Sansone, C., and Vento, M., An improved algorithm for matching large graphs, *Proc. IAPR GbRPR'01*, 149–159, 2001.

[8] Das, C., Raghothama, S., and Balaram, P., A designed three stranded beta-sheet peptide as a multiple beta-hairpin model, *J. Am. Chem. Soc.*, 120:5812–5813, 1998.

[9] Dror, O., Benyamini, H., Nussinov, R., and Wolfson, H., MASS: multiple structural alignment by secondary structures, *Bioinformatics*, 19(Suppl. 1):i95–i104, 2003.

[10] Efimov, A. V., Super-secondary structures involving triple-strand beta-sheets, *FEBS Lett.*, 334:253–256, 1993.

[11] Eidhammer, I., Jonassen, I., and Taylor, W. R., Protein structure comparison and structure patterns, *J. Comp. Biol.*, 7:685–716, 2000.

[12] Fernandez-Fuentes, N., Oliva, B., and Fiser, A., A supersecondary structure library and search algorithm for modeling loops in protein structures, *Nucleic Acids Res.*, 34:2085–2097, 2006.

[13] Frishman, D. and Argos, P., Knowledge-based secondary structure assignment, *Prot. Struct. Funct. Genet.*, 23:566–579, 1995.

[14] He, H. and Singh, A. K., GraphRank: statistical modeling and mining of significant subgraphs in the feature space, *Proc. ICDM'06*, 885–890, 2006.

[15] Huan, J., Bandyopadhyay, D., Prins, J., Snoeyink, J., Tropsha, A., and Wang, W., Distance-based identification of spatial motifs in proteins using constrained frequent subgraph mining, *Proc. CSB'06*, 227–238, 2006.

[16] Huan, J., Bandyopadhyay, D., Wang, W., Snoeyink, J., Prins, J., and Tropsha, A., Comparing graph representations of protein structure for mining family-specific residue-based packing motifs, *J. Comp. Biol.*, 12:657–671, 2005.

[17] Jonassen, I., Eidhammer, I., Conklin, D., and Taylor, W. R., Structure motif discovery and mining the PDB, *Bioinformatics*, 18:362–367, 2002.

[18] Kagawa, W., Kurumizaka, H., Ishitani, R., Fukai, S., Nureki, O., Shibata, T., and Yokoyama, S., Crystal structure of the homologous-pairing domain from the human Rad52 recombinase in the undecameric form, *Mol. Cell*, 10:359–371, 2002.

[19] Koch, I., Lengauer, T., and Wanke, E., An algorithm for finding maximal common subtopologies in a set of protein structures, *J. Comp. Biol.*, 3:289–306, 1996.

[20] Kurochkina, N. and Privalov, G., Heterogeneity of packing: structural approach, *Protein Sci.*, 7:897–905, 1998.

[21] Mitchell, E. M., Artymiuk, P. J., Rice, D. W., and Willett, P., Use of techniques derived from graph theory to compare secondary structure motifs in proteins, *J. Mol. Biol.*, 212:151–166, 1989.

[22] Pei, J., Jiang, D., and Zhang, A., On mining cross-graph quasi-cliques, *Proc. SIGKDD'05*, 228–238, 2005.

[23] Rao, S. T. and Rossman, M. G., Comparison of super-secondary structures in proteins, *J. Mol. Biol.*, 76:241–256, 1973.

[24] Sun, Z. and Blundell, T., The pattern of common supersecondary structure (motifs) in protein database, *Proc. HICSS'95*, 312–318, 1995.

[25] Szustakowski, J. D., Kasif, S., and Weng, Z., Less is more: towards an optimal universal description of protein folds, *Bioinformatics*, 21(Suppl. 2):ii66–ii71, 2005.

[26] Taylor, W. R. and Thornton, J. M., Prediction of super-secondary structure in proteins, *Nature*, 301:540–542, 1983.

[27] Wang, J., Zeng, Z., and Zhou, L., CLAN: an algorithm for mining closed cliques from large dense graph databases, *Proc. ICDE'06*, 73, 2006.

[28] Yan, X. and Han, J., CloseGraph: mining closed frequent graph patterns, *Proc. SIGKDD'03*, 286–295, 2003.

[29] http://astral.berkeley.edu/

[30] http://scop.mrc-lmb.cam.ac.uk/scop/

[31] http://softsurfer.com/Archive/algorithm_0106/algorithm_0106.htm

FragQA: predicting local fragment quality of a sequence-structure alignment

Xin Gao[1]   Dongbo Bu[1,3]
x4gao@cs.uwaterloo.ca   dbu@cs.uwaterloo.ca

Shuai Cheng Li[1]   Jinbo Xu[2]   Ming Li[1]
scli@cs.uwaterloo.ca   j3xu@tti-c.org *   mli@cs.uwaterloo.ca

[1] *David R. Cheriton School of Computer Science, University of Waterloo, Waterloo, ON, Canada, N2L 3G1*
[2] *Toyota Technological Institute at Chicago, Chicago, IL, USA, 60637*
[3] *Institute of Computing Technology, Chinese Academy of Sciences, Beijing, China, 100080*

Motivation.

Although protein structure prediction has made great progress in recent years, a protein model derived from automated prediction methods is subject to various errors. As methods for structure prediction develop, a continuing problem is how to evaluate the quality of a protein model, especially to identify some well predicted regions of the model, so that the structure biology community can benefit from automated structure prediction. It is also important to identify badly-predicted regions in a model so that some refinement measurements can be applied to.

Results.

We present a novel technique FragQA to accurately predict local quality of a sequence-structure (i.e., sequence-template) alignment generated by comparative modeling (i.e., homology modeling and threading). Different from previous local quality assessment methods, FragQA directly predicts cRMSD between a continuously aligned fragment determined by an alignment and the corresponding fragment in the native structure. FragQA uses an SVM (Support Vector Machines) regression method to perform prediction using information extracted from a single given alignment. Experimental results demonstrate that FragQA performs well on predicting local quality. More specifically, FragQA has prediction accuracy better than a top performer ProQres [18]. Our results indicate that (1) local quality can be predicted well; (2) local sequence evolutionary information (i.e., sequence similarity) is the major factor in predicting local quality; and (3) structure information such as solvent accessibility and secondary structure helps improving prediction performance.

Keywords: Local quality assessment; SVM regression; sequence-structure alignment.

1. Introduction

The biennial CASP (Critical Assessment of Structure Prediction) [12–15] events have demonstrated that the three-dimensional structures of many new target pro-

---

*To whom correspondence should be addressed.

teins can be predicted at a reasonable resolution, although in most cases, the predicted models are still not accurate enough for functional study. In particular, comparative modeling methods can generate reasonably good models for approximately 70% of target proteins in recent CASP events. Even for those FM (free modeling) targets, a structural model generated by protein threading usually contains some good local regions, although the overall conformation of the model is incorrect [21].

As methods for structure prediction develop, a continuing problem is how to evaluate the quality of a protein model in details. The challenge is to distinguish a good model from a bad one (as referred to global quality assessment) as well as correctly-predicted residues from badly-predicted ones (as referred to local quality assessment). To make automated structure prediction really useful for the structure biology community, a reliable model quality evaluation program is indispensable when hundreds of models are predicted for a single target protein. There are a variety of global quality prediction methods [3, 5, 10, 17, 19]. This kind of programs can be used to pick up the best few from a bunch of models generated by different structure prediction programs, which enables structure biologists to focus on the most possible models. In addition, a common practice taken by some human predictors or consensus-based automatic predictors to further improve the accuracy of structure prediction is to identify correctly-predicted regions from each structural model and then assemble them together to obtain a better overall model for the target protein; for example, 3D-SHOTGUN [4] and TASSER [21] are two such top-scoring methods. This kind of refinement methods often perform better than the classical threading-based protein structure prediction methods. The key factor underlying the success of these refinement methods is identifying the correctly-predicted regions in a structural model. Besides being used to examine and improve the accuracy of a protein model, local quality prediction methods can also be used to recognize functional residues in a protein model [1, 16].

Local quality assessment methods are either structure-based or alignment-based. ERRAT [2] is a program that uses only structure information. This program employs a Gaussian error function based on the statistics of non-bonded interactions to predict incorrect regions in a protein model. These methods can recognize incorrect structural regions which obviously deviate from their natives. There are also some programs using alignment information to predict local quality. Tress *et al* developed a method to evaluate local quality of a given alignment and tested the method on alignments generated by five comparative modeling methods [16]. The results indicate that an alignment position with high profile-derived alignment score often has good quality. Wallner *et al* developed four neural network-based methods [18] to identify correct regions in a protein model, using either structure information or alignment information: ProQres, ProQprof, ProQlocal and Pcons-local. ProQres uses structure information in a protein model; while ProQprof uses alignment information such as profile-profile scores, information scores, and gap penalty. ProQlocal combines ProQres and ProQprof together to achieve a better performance. Pcons-local is a consensus-based local quality predictor, taking as input protein models generated by different structure prediction programs.

Our contribution. In this paper, we present a novel method FragQA to accurately predict local quality of a sequence-structure alignment. Distinguishing itself from its peers, FragQA predicts the quality of an ungapped region (referred to as fragment) in the alignment. The quality is measured using the cRMSD (i.e., C_α-based RMSD) between two fragments corresponding to the ungapped region: one is the native structure of the region and the other is the predicted structure. Furthermore, statistical significance is introduced to improve FragQA's performance. As opposed to cRMSD, statistical significance can cancel out the impact of region length. FragQA utilizes only information in a single alignment. Structure information in the alignment-derived protein model is not directly used. However, in calculating features from an alignment, we use structure information in the template.

2. Methods

2.1. *Problem description*

This paper studies the following problem: Given a sequence-structure alignment, what is the quality of an ungapped region in this alignment? The quality is defined as the cRMSD between the native and the predicted local conformations of the ungapped region, denoted as "cRMSD of an ungapped region", after they are optimally superimposed. Please note that the two conformations are superimposed without taking into consideration other parts of the alignment. The reason to do this local superimposition is to eliminate the impact by some badly predicted regions of the model, and evaluate how truly similar a region in a model is to the native one. The alignment is cut into ungapped regions at gap positions.

2.2. *Development of FragQA*

Our SVM regression model uses only features extracted from a single sequence-template alignment, generated by any threading program. To exploit the evolutionary information of proteins, we utilize sequence profile of both target protein and template protein in calculating features. The sequence profile of the template, denoted by $PSSM_{template}$ (position specific mutation matrix), is generated by PSI-BLAST with five iterations; $PSSM_{template}(i, a)$ encodes mutation information for amino acid a at position i of the template. We also apply PSI-BLAST with five iterations to generate position specific frequency matrix, $PSFM_{target}$, for each target protein; $PSFM_{target}(j, b)$ encodes occurring frequency of amino acid b at position j of the target. Let $A(i)$ denote the aligned sequence position of template position i, and T_{temp} denote the set of template positions belonging to an aligned region. We studied a variety of features extracted from the alignment and later we will discuss their relative importance. In summary, we tested the following features in FragQA:

(1) *Mutation score*: Mutation score measures the sequence similarity between two segments of an aligned region: one corresponds to the target protein and the

other to the template. The mutation score (S_m) of a region is calculated as:

$$S_m = \sum_{i \in T_{temp}} \sum_{a} PSFM_{target}(A(i), a) \times PSSM_{template}(i, a) \qquad (1)$$

(2) *Environmental fitness score*: This score measures how well to align one target protein region to the environment where the template protein region lies in. The environment consists of two types of local structure features.

- Three types of secondary structure are used: α-helix, β-strand, and loop.
- Solvent accessibility: There are three levels: buried (inaccessible), intermediate, and accessible. The Equal-Frequency discretization method is used to determine boundaries between these three levels. The calculated boundaries are 7% and 37%.

Thus, there are nine environment combinations (denoted as env) in total. Let $F(env, a)$ denote the environment fitness potential for amino acid a and environment combination env, which is taken from PROSPECT-II [9]. The environment fitness score (S_e) for an aligned region is calculated as:

$$S_e = \sum_{i \in T_{temp}} \sum_{a} PSFM_{target}(A(i), a) \times F(env_i, a) \qquad (2)$$

(3) *Secondary structure score*: In addition to secondary structure information encoded in environmental fitness score, we also use $SS(i, A(i))$, the secondary structure difference between position i in template and position $A(i)$ in target, to measure the quality of an ungapped region from another aspect. We use PSIPRED [7] to predict the secondary structure of the target protein. Let $\alpha(j)$, $\beta(j)$ and $loop(j)$ denote the predicted confidence levels of α-helix, β-sheet and loop at sequence position j, respectively. If the secondary structure type at template position i is α-helix, then $SS(i, A(i)) = \alpha(A(i)) - loop(A(i))$. If the secondary structure type at template position i is β-sheet, then $SS(i, A(i)) = \beta(A(i)) - loop(A(i))$. Otherwise, we set $SS(i, A(i))$ to be 0. The secondary structure score (S_{ss}) of an ungapped region is calculated as:

$$S_{ss} = \sum_{i \in T_{temp}} SS(i, A(i)) \qquad (3)$$

(4) *Contact capacity score*: Contact capacity potentials describe the hydrophobic contribution of free energy, measured by the capability of a residue making a certain number of contacts with other residues in the protein. Two residues are in physical contact if the spatial distance between their C_β atoms (C_α for glycine) is smaller than 8Å. Let $CC(a, k)$ denote the contact potential of amino acid a having k contacts. $CC(a, k)$ is calculated by statistics on PDB as:

$$CC(a, k) = -log \frac{N(a, k)N}{N(k)N'(a)} \qquad (4)$$

where $N(a, k)$ is the number of amino acid a with k contacts; $N(k)$ is the number of residues with k contacts; $N'(a)$ is the number of amino acid a; and N is the total number of residues in PDB. Let $C(i)$ denote the number of contacts at template position i. The contact capacity score (S_c) is calculated as:

$$S_c = \sum_{i \in T_{temp}} \sum_a PSFM_{target}(A(i), a) \times CC(a, C(i)) \tag{5}$$

(5) *Aligned region length*: The cRMSD between two fragments of an ungapped region is relevant to its length. The longer the ungapped region is, the more likely larger the cRMSD is.

(6) *Z-score*: Z-score measures the overall quality of a sequence-structure alignment. An alignment with a good Z-score likely contains more good ungapped regions. In this paper, Z-score is predicted alignment accuracy normalized by target protein size, and calculated by Xu's SVM module [19].

(7) *Alignment topology*: We test 3 separate topology features: template protein size, target protein size, alignment length (i.e., the number of aligned positions).

(8) *Sequence identity*: We use the fraction of identical residues in the whole alignment to measure the sequence identity.

Meanwhile, feature (1)-(5) are specific to the ungapped region; while feature (6)-(8) are for the whole sequence-structure alignment.

3. Results

3.1. *FragQA Training*

Training and Test Data. Choosing good training and test sets is one of the key steps in objectively evaluating the performance of a machine learning method. We test our method on several threading methods, such as RAPTOR [20] (with three different threading algorithms), PROSPECT-II [9], and GenTHREADER [8]. The results are similar. In this paper, we only show the results on alignments generated by RAPTOR default threading algorithm (with NoCore option). Our training and test data is from recent CASP7 event. There are 104 target proteins in CASP7 while only 92 of them have native structures published after the event. Ninety-one target proteins are left after we removed redundancy at 40% sequence identity level using CD-HIT [11]. Only T0346 is removed because it shares 71% sequence identity with T0290. To do a cross validation, the 91 target proteins are randomly divided into four sets. Here, we took top 10 alignments generated by RAPTOR for each target protein. If one target protein belongs to a set, then all of its 10 alignments belong to this set. Each alignment is cut into a set of ungapped regions with cutting points being at the gap positions. The ungapped regions containing less than 5 residues are not considered in our experiments. Table 1 shows the statistics on the four sets. It is clear that the four data sets are very similar.

Training. We used the software SVM-light [6] with RBF (radial basis function) kernel to train FragQA. The parameter gamma in the RBF kernel function is trained using the leave-one-out error estimation method. Other parameters are set to their default values or calculated automatically by SVM-light. Experimental results indicate that the RBF kernel with its gamma parameter set to 0.2 can yield the best

Table 1.   Statistics on the four data sets. Column 2-5 show the number of target proteins, the number of fragments, the average cRMSD of the fragments, and the standard deviation of cRMSD of each set, respectively.

Set Name	# of proteins	# of fragments	Average cRMSD	Deviation
1	23	1347	2.93Å	1.50Å
2	22	1108	2.57Å	1.46Å
3	23	1519	2.86Å	1.47Å
4	23	1461	2.73Å	1.49Å

training performance. Other kernel functions such as linear kernel and polynomial kernel are also tested, but they cannot yield as good performance as the RBF kernel.

We executed a 4-fold cross validation. Each time we used three of the four data sets as the training set, and the other one for testing.

3.2. Performance of FragQA

After studying the relative importance of the 8 features, which will be discussed later, we encoded following features into FragQA: (1) length of the ungapped region; (2) Z-score of the whole alignment; (3) mutation score of the region; (4) environmental fitness score of the region; and (5) secondary structure score of the region.

3.2.1. Comparing to ProQres

As far as we know, FragQA is the first method to directly predict the local fragment quality. Thus, there is no existing method for us to compare with. However, there are some well-known methods that predict local quality for each residue. So it is possible to convert the prediction on residues by such methods to a prediction of a fragment. Since the objective function of FragQA is cRMSD, to fairly evaluate FragQA, we compared FragQA to a top-notch method ProQres [18], which uses a residue-based cRMSD-related objective function. We tested all three available methods by ProQgroup in terms of the ability to predict fragment quality : ProQlocal, ProQres, and ProQprof. ProQres yielded the best results (slightly better than ProQlocal and ProQprof in terms of fragment cRMSD prediction). Thus, in this paper, we will compare FragQA to ProQres. The objective function of ProQres is $D_i = 1/(1 + \frac{(d_i)^2}{(d_0)^2})$ [18], where d_i denotes the cRMSD at position i, and d_0 is set to $\sqrt{5}$. From the prediction of ProQres, we can calculate d_i from D_i for each residue of a fragment, then use $cRMSD = \sqrt{\frac{1}{n}\sum_{i=1}^{n} d_i^2}$ to compute the predicted cRMSD by ProQres for the fragment, where n is the length of the fragment. Note cRMSD calculated by this way has a slightly different meaning to the one used by FragQA, because this cRMSD is based on the optimal superposition between the whole target and the template on all similar regions, while FragQA's cRMSD is based on the optimal superposition between two fixed regions. However, the superposition between two aligned regions determined by the optimal superposition of the whole target and template is usually very similar to the optimal one between the two regions, because aligned regions are usually very similar. Thus, FragQA and ProQres are comparable from this point of view.

3.2.2. *Prediction Error and Correlation Coefficient of FragQA*

The prediction error is defined as the difference between the predicted cRMSD values and the real ones. Table 2 lists the average prediction errors of FragQA and ProQres, under different cRMSD thresholds on the four test sets, together with average fraction of fragments with real cRMSD under such thresholds, and the correlation coefficient between the predicted and real cRMSD by FragQA and ProQres on the four test sets. As shown in this table, the prediction error of FragQA ranges from 0.9Å to 1.6Å, while the error of ProQres ranges from 0.9Å to 2.4Å. In most cases, the prediction error of FragQA is much smaller than that of ProQres. In fact, when there is no restriction on cRMSD, the error of FragQA is on average 0.5Å smaller than that of ProQres. The smallest error of FragQA happens when cRMSD threshold is set to 3Å, which means FragQA is most accurate when dealing with fragments with cRMSD to native smaller than 3Å. However, when the real cRMSD is very small (≤ 1Å), the prediction error tends to be big. In other word, it is hard to obtain an accurate prediction when cRMSD is very small. As indicated in Table 2, the correlation coefficient between predicted cRMSD by FragQA and the real cRMSD is about 0.5 for each test set, while that of ProQres is at most 0.22.

Table 2. The prediction error of FragQA (denoted as FQA) and ProQres (denoted as PQr), under different cRMSD thresholds on the four test sets, average fraction of fragments with real cRMSD under such thresholds, and the correlation coefficient of FragQA and ProQres.

cRMSD	Test Set 1		Test Set 2		Test Set 3		Test Set 4		Ave. Fraction
	FQA	PQr	FQA	PQr	FQA	PQr	FQA	PQr	
$\leq$1Å	1.36	1.50	1.57	1.10	1.41	1.35	1.54	1.30	14%
$\leq$2Å	1.11	1.06	1.28	0.90	1.08	1.01	1.18	1.01	42%
$\leq$3Å	1.00	1.84	1.16	1.12	0.94	0.98	1.04	1.01	69%
$\leq$4Å	1.03	1.79	1.12	1.23	0.97	1.21	1.04	1.13	85%
$\leq$5Å	1.12	1.88	1.14	1.34	1.06	1.37	1.09	1.34	92%
$\leq$6Å	1.20	1.98	1.19	1.46	1.16	1.50	1.20	1.51	95%
$\leq$7Å	1.33	2.13	1.26	1.57	1.22	1.58	1.25	1.62	97%
$\leq$8Å	1.41	2.20	1.32	1.68	1.29	1.67	1.31	1.72	98%
$\leq$9Å	1.48	2.27	1.36	1.73	1.37	1.78	1.36	1.77	99%
$\leq$10Å	1.57	2.37	1.39	1.77	1.41	1.84	1.41	1.83	99%
Correlation Coefficient	0.51	0.07	0.46	0.22	0.50	0.22	0.48	0.16	-

3.2.3. *Sensitivity and Specificity*

Given a cRMSD threshold, sensitivity is calculated as the fraction of ungapped regions with real cRMSD smaller than the threshold, that are also predicted to be smaller than the threshold. Specificity measures the fraction of ungapped regions with predicted cRMSD under a given threshold, that indeed have cRMSD smaller than the threshold. Figure 1 illustrates the sensitivity and specificity of FragQA and ProQres under various cRMSD thresholds on the four test sets.

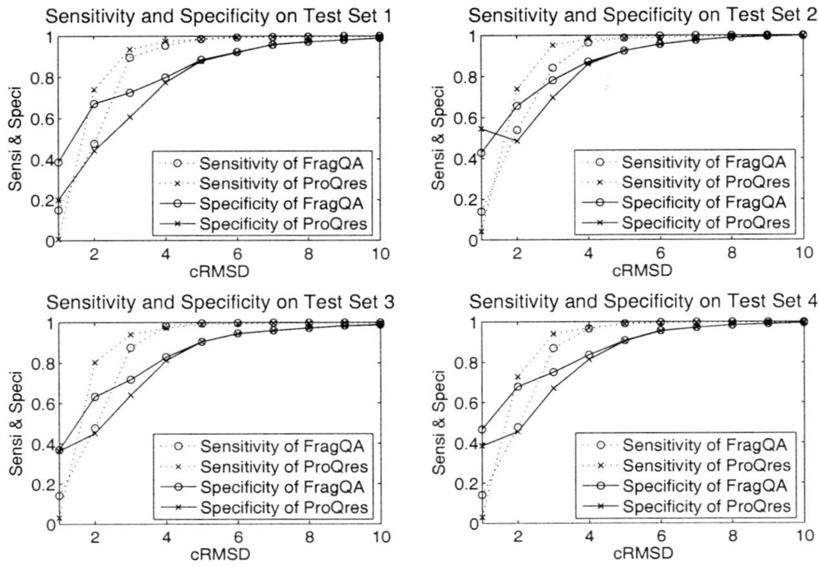

Fig. 1.   Comparison of sensitivity and specificity between FragQA and ProQres under different cRMSD thresholds on the four test sets. Circle dotted line: sensitivity of FragQA, cross dotted line: sensitivity of ProQres, circle solid line: specificity of FragQA, cross solid line: specificity of ProQres. Please see Section 3.2.3 for the definition of sensitivity and specificity.

As shown in Figure 1, there is no obvious difference between sensitivity or specificity of FragQA and ProQres when cRMSD is larger than 4Å. When cRMSD is smaller than 4Å, the sensitivity of ProQres is higher than that of FragQA for most cases, while the specificity of FragQA is higher than that of ProQres. In particular, when cRMSD threshold is 2.5Å, approximately 70% of ungapped regions with predicted cRMSD by FragQA under 2.5Å indeed have cRMSD less than 2.5Å, while 70% of ungapped regions with real cRMSD under this threshold are predicted by FragQA correctly. However, the sensitivity of ProQres on cRMSD 2.5Å is about 80%, while the specificity is only 50%. This implies that ProQres has a strong trend to predict the cRMSD of a fragment to be smaller than the real value. This makes ProQres to have a high sensitivity while having a low specificity. As shown in Figure 1, the specificity curve of ProQres is not smooth sometime, nor it is monotonous. It is clear that for both FragQA and ProQres, the sensitivity curve increases much more quickly than the specificity curve. However, all curves are quite low when cRMSD is small. A possible explanation is that when the ungapped region is short, a small cRMSD does not necessarily mean that this region has good quality. Therefore, it is hard for FragQA to predict cRMSD accurately under such cases. Later in this paper, we will replace cRMSD with its statistical significance and show that when statistical significance is high, even as high as 1, FragQA can still yield a good prediction.

3.2.4. *Feature Selection for FragQA*

It is important to detect which features are closely relevant to the prediction capability of FragQA since unrelated features may introduce extra noise. We studied the importance of each feature by excluding it from the feature set, training a new FragQA, and then testing the performance of this new predictor. Thus, we can compare the performance resulting from different sets of features and then detect the important features.

Table 3. Sensitivity of FragQA with different feature sets. The 2nd column lists the sensitivity of FragQA with all features. Starting from the 3rd column, each column lists the sensitivity when one feature is removed. *Len*: region length, S_z: Z-score, S_m: mutation score, S_e: environmental fitness score, S_c: contact capacity score, S_{ss}: secondary structure score, *Topo*: topology features, *SeqId*: sequence identity.

cRMSD	All	No *Len*	No S_z	No S_m	No S_e	No S_c	No S_{ss}	No *Topo*	No *SeqId*
$\leq$1Å	0.12	0	0.04	0.09	0.11	0.13	0.13	0.12	0.12
$\leq$1.25Å	0.16	0.01	0.08	0.15	0.14	0.22	0.18	0.15	0.16
$\leq$1.5Å	0.25	0.04	0.16	0.19	0.22	0.27	0.26	0.25	0.25
$\leq$1.75Å	0.35	0.12	0.27	0.27	0.29	0.34	0.36	0.35	0.34
$\leq$2Å	0.42	0.21	0.38	0.35	0.39	0.48	0.42	0.42	0.43
$\leq$2.25Å	0.50	0.42	0.52	0.46	0.48	0.58	0.51	0.51	0.51
$\leq$2.5Å	0.62	0.61	0.64	0.55	0.56	0.65	0.63	0.62	0.62
$\leq$2.75Å	0.70	0.74	0.73	0.65	0.67	0.74	0.71	0.69	0.69
$\leq$3Å	0.76	0.82	0.79	0.74	0.75	0.81	0.77	0.76	0.76
$\leq$3.25Å	0.83	0.90	0.86	0.82	0.80	0.85	0.84	0.83	0.83
$\leq$3.5Å	0.88	0.94	0.90	0.88	0.84	0.89	0.89	0.88	0.88

Table 4. Specificity of FragQA with different feature sets. The 2nd column lists the specificity of FragQA with all features. Starting from the 3rd column, each column lists the specificity when one feature is removed. *Len*: region length, S_z: Z-score, S_m: mutation score, S_e: environmental fitness score, S_c: contact capacity score, S_{ss}: secondary structure score, *Topo*: topology features, *SeqId*: sequence identity.

cRMSD	All	No *Len*	No S_z	No S_m	No S_e	No S_c	No S_{ss}	No *Topo*	No *SeqId*
$\leq$1Å	0.19	0	0.10	0.17	0.16	0.32	0.17	0.18	0.18
$\leq$1.25Å	0.28	0.22	0.20	0.27	0.22	0.43	0.27	0.28	0.28
$\leq$1.5Å	0.42	0.23	0.37	0.35	0.36	0.49	0.41	0.41	0.42
$\leq$1.75Å	0.52	0.41	0.51	0.46	0.47	0.57	0.51	0.52	0.52
$\leq$2Å	0.59	0.48	0.58	0.53	0.57	0.65	0.56	0.59	0.60
$\leq$2.25Å	0.64	0.56	0.64	0.60	0.62	0.68	0.63	0.64	0.64
$\leq$2.5Å	0.72	0.63	0.70	0.66	0.69	0.73	0.70	0.72	0.72
$\leq$2.75Å	0.78	0.67	0.75	0.73	0.76	0.78	0.77	0.77	0.78
$\leq$3Å	0.79	0.70	0.77	0.77	0.80	0.79	0.79	0.79	0.79
$\leq$3.25Å	0.82	0.75	0.80	0.81	0.83	0.82	0.80	0.82	0.82
$\leq$3.5Å	0.86	0.79	0.84	0.83	0.85	0.86	0.84	0.86	0.86

Tables 3 and 4 list the sensitivity and specificity of FragQA with different sets of features under different cRMSD thresholds on test set 1. The results are similar on other test sets. There is no obvious difference among different sets of features when cRMSD threshold is larger than 3.75Å. As shown in these two tables, if we remove the aligned region length, the performance of FragQA will drop obviously, except for cRMSD threshold larger than 2.75Å, the sensitivity of FragQA without fragment length is a little higher than that with all features. This complies with a fact that cRMSD itself is closely related to the length of an ungapped region. Removing mutation score or the overall Z-score will also have an obvious reduction on the performance of FragQA, except for cRMSD larger than 2.25Å, removing Z-score will increase sensitivity slightly and have no obvious influence on specificity. This also makes sense: mutation score measures the sequence similarity in the aligned region and Z-score evaluates the overall quality of the alignment. An alignment with good overall quality often contains good aligned regions. However, when the overall quality of an alignment is poor (Z-score is low), the fragments can be either good or bad. In such case, Z-score will not be an influential factor any more. Removing environmental fitness score will decrease both the sensitivity and specificity. Surprisingly, removing contact capacity score will increase both sensitivity and specificity. This implies contact score is a noise feature. On the other hand, removing secondary structure score will decrease the specificity but increase the sensitivity slightly. Removing any other features, such as alignment topology features and sequence identity feature, does not obviously deteriorate either sensitivity or specificity. Thus, the final version of FragQA uses the following features: (1) aligned region length; (2) overall alignment Z-score; (3) mutation score; (4) environmental fitness score; and (5) secondary structure score. Meanwhile, mutation score, Z-score and the region length are the most important factors in quality prediction.

3.2.5. *Statistical Significance*

The cRMSD between the predicted structure of an ungapped region and its native is closely relevant to the length of the region. Thus, a 5-residue ungapped region with 3Å cRMSD may not be better than a 15-residue region with 4Å cRMSD. To better evaluate the quality of a region, we calculate the statistical significance of its cRMSD to reduce the bias introduced by region length. To calculate statistical significance, statistical distribution of cRMSD for a given region length is empirically calculated as follows. For a given region length, we randomly sampled 10,000 pairs of fragments of this length from PDB30 and then calculated their cRMSDs. PDB30 is a subset of PDB, in which any two proteins share no more than 30% sequence identity. As shown in Figure 2(a), the mean of cRMSD increases clearly with respect to the length, but the standard deviation increases much more slowly. The cRMSD distribution looks like a normal distribution (figure not shown). For a given ungapped region with length l and (real or predicted) cRMSD r, its statistical significance (denoted as *StatSig*) is calculated as follows:

$$StatSig = \frac{\#random\ pairs\ of\ length\ l\ with\ cRMSD \geq r}{10,000} \tag{6}$$

Thus, the smaller the cRMSD is, the larger its statistical significance is.

We calculated the sensitivity and specificity of FragQA in terms of statistical significance in a way similar to that calculates them in terms of cRMSD. For each statistical significance threshold varying from 0 to 1, the sensitivity is defined as the percentage of ungapped regions with real statistical significance larger or equal than the threshold, that also have predicted values larger or equal than the threshold. The specificity is defined as the percentage of ungapped regions with predicted significance larger or equal than the threshold, that have real statistical significance better or equal than the threshold. Figure 2(b) illustrates the sensitivity and specificity of FragQA in terms of statistical significance on test set 1. Results are similar on other three sets. As shown in this figure, when statistical significance is 0.8 (about 81% fragments in our test sets have such values), both the sensitivity and specificity is around 90%. Even when statistical significance threshold is 1 (about 48% fragments in our test sets have this value), the sensitivity is 78%, and the specificity is 88%.

We also studied the prediction error and correlation coefficient of FragQA in terms of statistical significance. The prediction error increases from 0.02 to 0.16 when the statistical significance threshold decreases from 1 to 0. Recall that it is hard for FragQA to predict cRMSD accurately when cRMSD is small, this result implies that FragQA is able to predict statistical significance well on high-quality fragments. On the other hand, the correlation coefficient of FragQA on each set in terms of statistical significance is higher than 0.60. This means statistical significance is probably a better way to measure quality of a fragment. We further compared FragQA to ProQres in terms of statistical significance. FragQA also outperforms

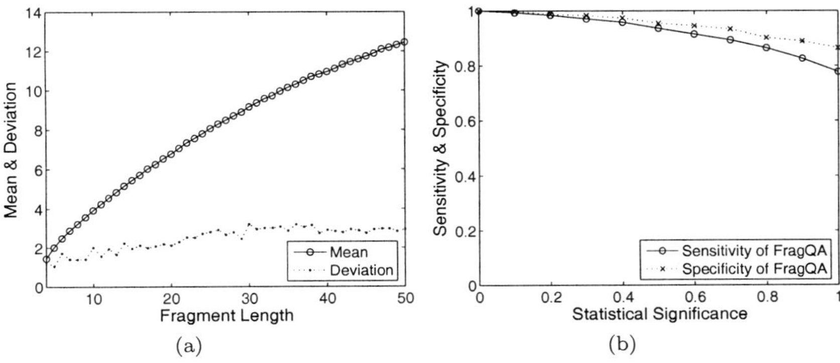

Fig. 2.   (a) Mean (circle solid line) and standard deviation (point dotted line) of cRMSD for random region sets with length from 5 residues to 50 residues. (b) FragQA's sensitivity (circle solid line) and specificity (cross dotted line) in terms of statistical significance on test set 1. Please see Section 3.2.5 for the definitions of sensitivity and specificity.

ProQres on prediction error, correlation coefficient, and specificity, but is not as good as ProQres on sensitivity (data not shown here).

4. Discussion

To the best of our knowledge, FragQA is the first program that directly predicts the quality of an ungapped region in an alignment. Currently FragQA utilizes only alignment information in a single alignment, although some structure information from the template is also taken into consideration. We plan to further improve the method along the following avenues: 1) Combine structure information in a protein model with alignment information; and 2) Utilize various alignments generated by independent threading programs so that consensus information can be used to boost prediction performance.

Although our experiments used alignments generated by RAPTOR as data source, FragQA can take as input alignments generated by any comparative modeling methods, since FragQA is totally independent of threading methods. We benchmarked FragQA using RAPTOR's results in CASP7 because CASP7 target protein structures were published only recently. Most CASP7 target proteins have low sequence similarity with proteins in PDB. The template database used by RAPTOR for CASP7 was generated before any CASP7 target structures were deposited into PDB. This can reduce bias introduced by template database to its minimum level.

A potential application of FragQA is to identify high-quality regions in an alignment. These regions can often cover a large portion of the target protein even if it is a hard target and thus, they can be re-assembled to obtain a better overall structural model. For example, Zhang-server [21] is the best automated server in CASP7. It first cuts a threading-generated alignment into some ungapped regions and then rearranges the physical orientations of these regions. Zhang-server uses all the ungapped regions without considering their quality. A further improvement over Zhang's method is to first predict the quality of each region and then refold only those high-quality regions to obtain a better structural model.

5. Conclusion

This research develops a novel local quality predictor, FragQA. Experimental results on the CASP7 data set demonstrate that FragQA performs well, better than a top-notch predictor ProQres. Our experimental results indicate that local sequence evolutionary information is the major factor in predicting local quality. Other information such as secondary structure and solvent accessibility also helps improving prediction accuracy.

Acknowledgment

This work is supported by NSERC Grant OGP0046506, and NSF of China Grant 60496324.

References

[1] Berezin, C., Glaser, F., Rosenberg, J., Paz, I., Pupko, T., Fariselli, P., Casadio, R., and Ben-Tal, N., ConSeq: the identification of functionally and structurally important residues in protein sequences, *Bioinformatics*, 8:1322–1324, 2004.

[2] Colovos, C., and Yeates, T.O., Verification of protein structures: patterns of non-bonded atomic interactions, *Proteins Science* 2:1511-1519, 1993.

[3] Eisenberg, D., Lthy, R., and Bowie, J.U., VERIFY3D: assessment of protein models with three-dimensional profiles, *Methods Enzymol*, 277:396–404, 1997.

[4] Fischer, D., 3D-SHOTGUN: A novel, cooperative, fold-recognition meta-predictor, *Proteins: Structure, Function and Genetics*, 51:434–441, 2003.

[5] Ginalski, K., Elofsson, A., Fischer, D., and Rychlewski, L., 3D-Jury: a simple approach to improve protein structure predictions, *Bioinformatics*, 19:1015–1018, 2003.

[6] Joachims, T., *Making large-scale SVM learning practical. Advances in Kernel Methods - Support Vector Learning*, MIT-Press, 1999.

[7] Jones, D.T., Protein secondary structure prediction based on position-specific scoring matrices, *Journal of Molecular Biology*, 292(2):195–202, 1999.

[8] Jones, D.T., GenTHREADER: an efficient and reliable protein fold recognition method for genomic sequences, *Journal of Molecular Biology*, 287:797–815, 1999.

[9] Kim, D., Xu, D., Guo, J., Ellrott, K., and Xu, Y., PROSPECT II: protein structure prediction method for genome-scale applications, *Protein Engineering* 16(9):641–650, 2003.

[10] Laskowski, R.A., Macarthur, M.W., Moss, D.S., and Thornton, J.M., PROCHECK: a program to check the stereochemical quality of protein structures, *Journal of Applied Crystallography*, 26(2):283–291, 1993.

[11] Li, W. and Godzik, A., CD-HIT: a fast program for clustering and comparing large sets of protein or nucleotide sequences, *Bioinformatics*, 22:1658–1659, 2006.

[12] Moult, J., Hubbard, T., Fidelis, K., and Pedersen, J., Critical assessment of methods of protein structure prediction (CASP):round III, *Proteins*, 37:2–6, 1999.

[13] Moult, J., Fidelis, K., Zemla, A., and Hubbard, T., Critical assessment of methods of protein structure prediction (CASP):round IV, *Proteins*, 45:2–7, 2001.

[14] Moult, J., Fidelis, K., Zemla, A., and Hubbard, T., Critical assessment of methods of protein structure prediction (CASP):round V, *Proteins*, 53:334–339, 2003.

[15] Moult, J., Fidelis, K., Rost, B., Hubbard, T. and Tramontano, A., Critical assessment of methods of protein structure prediction (CASP):round 6, *Proteins*, 61:3–7, 2005.

[16] Tress, M., Jones, D., and Valencia, A., Predicting reliable regions in protein alignments from sequence profiles, *Journal of Molecular Biology*, 330:705–718, 2003.

[17] Wallner, B., and Elofsson, A., Can correct protein models be identified? *Protein Science*, 12(5):1073–1086, 2003.

[18] Wallner, B., and Elofsson, A., Identification of correct regions in protein models using structural, alignment, and consensus information, *Protein Science*, 15:900–913, 2005.

[19] Xu, J., Protein fold recognition by predicted alignment accuracy, *ACM/IEEE Transactions on Computational Biology and Bioinformatics*, 2(2):157–165, 2005.

[20] Xu, J., Li, M., Kim, D., and Xu, Y., RAPTOR: optimal protein threading by linear programming, *JBCB*, 1(1):95–117, 2003.

[21] Zhang, Y., and Skolnick, J., Automated structure prediction of weakly homologous proteins on a genomic scale, *Proceedings of National Academy of Sciences*, 101(20):7594–7599, 2004.

PREDICTING B CELL EPITOPE RESIDUES WITH NETWORK TOPOLOGY BASED AMINO ACID INDICES

JIAN HUANG[1,2]
hjian@kuicr.kyoto-u.ac.jp

WATARU HONDA[1]
honda@kuicr.kyoto-u.ac.jp

MINORU KANEHISA[1,3]
kanehisa@kuicr.kyoto-u.ac.jp

[1] *Bioinformatics Center, Institute for Chemical Research, Kyoto University, Gokasho Uji, Kyoto 611-0011, Japan*
[2] *School of Life Science and Technology, University of Electronic Science and Technology of China, Chengdu 610054, China*
[3] *Human Genome Center, Institute of Medical Science, University of Tokyo, 4-6-1 Shirokanedai, Minato-ku, Tokyo 108-8639, Japan*

We evaluate the performance of six amino acid indices in B cell epitope residue prediction using the classical sliding window method on five data sets. Four of the indices: *i.e.* relative connectivity, clustering coefficient, closeness and betweenness are newly derived from the topological parameters of residue networks. The other two are Parker's hydrophilicity and Levitt's index, known as the best indices so far for B cell epitope prediction. On four of the data sets, the performance of all the indices was comparable and poor in general. When applied to one well-annotated data set, the performances improved and the 4 network based indices showed better performance than that of Parker's hydrophilicity and Levitt's index. When using the relative connectivity index on this data set, the prediction accuracy, sensitivity and specificity reached 73.6%, 73.0% and 75.0% respectively, with an area under the curve about 0.796. Thus, we suggested that this index is a good choice for B cell epitope prediction. It also indicates that the low performance of B cell epitope prediction is not only due to the methods and amino acid indices used, but also the data set as well. Interestingly, on the well-annotated data set, the performance of B cell epitope residue prediction is very similar to that of protein surface residue prediction, especially at the 10 and 20 Å^2 cutoffs. It is suggested that the performance in surface residue prediction might form a theoretical upper limit for the performance of B cell epitope residue prediction methods.

Keywords: B cell, epitope prediction, amino acid index, network topology

1. Introduction

The B cell epitopes of proteins are special regions on proteins that can be recognized by the antigen binding sites of antibodies or B cell receptors. Identified B cell epitopes are very useful because they can further be developed into diagnostics [1], therapeutics and vaccines [2, 3]. Therefore, it's only natural that B cell epitope mapping has been a major field of immunology research.

As identifying B cell epitopes experimentally is time-consuming and expensive, techniques to predict B cell epitopes have been developed for almost 30 years [4-17]. Most of these techniques are sliding window based sequence profiling methods. In brief, a window slides from the N-terminal to C-terminal of the query protein sequence. The mean propensity value of the residues in the window is then assigned to the residue in the middle according on the amino acid index [18-20] (also known as the propensity scale) used in the prediction. By combining such predictions with experimental verification, many successful cases have been reported. However, the performance of this kind of B cell epitope predictions has been disputed [21-23]. In a recent report, Blythe *et al* assessed 484 amino acid indices in the AAindex database [20] with sequence profiling methods. They found that even the best set of amino acid indices performed only marginally better than random [24], indicating that better methods or new amino acid indices are needed for B cell epitope prediction. A very recent study has confirmed that Parker's hydrophilicity (*Ph*) [8] and Levitt's index (*Li*) [12] are the best two indices so far for sequence profiling based B cell epitope prediction. However, even the performance of *Ph* and *Li* are unsatisfactory [25].

In a previous study, we built four new amino acid indices, termed relative connectivity (*Rk*), relative clustering coefficient (*Rc*), relative closeness (*Ro*) and relative betweenness (*Rb*), based on residue networks constructed from 640 representative PDB structures [26]. Compared with *Ph* and *Li*, these network topology based indices have shown better performance in protein surface residue prediction [26]. Surface residue prediction is related to B cell epitope prediction, due to the requirement for epitopes to be surface accessible to interact with an antibody [27, 28]. Since the network topology based indices have significantly better performance than *Ph* and *Li* in protein surface residue prediction [26], will they perform better in B cell epitope residue prediction too?

To answer the above question, the performance of *Ph*, *Li* and the 4 residue network topology derived amino acid indices in B cell epitope residue prediction are evaluated and compared in this study.

2. Methods and Data Sets

2.1. *Data sets*

Five data sets of proteins with annotated B cell epitope residues are used in the current study. The first data set, originally composed by Pellequer *et al*, contains 14 protein sequences and 82 epitopes [15]. We took the recreated electronic form of this dataset from the Lund group [25]. The second and the third data sets, denoted as AntiJen data set and HIV data set respectively, are composed by the Lund group [25]. The fourth data set is the DiscoTope data set, which has 75 antigens with B cell epitope residue annotation [29].

We compiled a fifth data set from the CED database [30]. Taking the well-studied hen egg white lysozyme (HEL) as the model antigen, 19 epitopes were found and used as annotation. We call this data set the HEL data set. For the purpose of comparison, the solvent accessible area of hen egg white lysozyme is computed from its PDB structure (1HEL) with the NACCESS program [31] using default parameters. Surface residues are assigned based on the solvent accessible area at different cutoff values of 1, 10, 20, 50 and 100 Å^2.

2.2. Amino acid indices

The 4 network topology based amino acid indices are taken from our pervious study [26]. The best two indices (Parker's hydrophilicity and Levitt's index) known for B cell epitope prediction are taken from references [12, 25]. They are listed in Table 1.

Table 1. Amino acid indices used in prediction.

	A	C	D	E	F	G	H	I	K	L	M	N	P	Q	R	S	T	V	W	Y
Rk	1.05	1.17	0.88	0.85	1.07	0.99	0.99	1.11	0.88	1.07	1.04	0.93	0.92	0.93	0.94	0.96	0.99	1.12	1.05	1.05
Rc	0.99	0.89	1.11	1.13	0.92	1.08	1.00	0.89	1.10	0.92	0.95	1.07	1.01	1.06	1.04	1.05	1.01	0.90	0.93	0.94
Ro	1.00	1.13	0.95	0.95	1.03	0.99	1.01	1.04	0.96	1.02	1.02	0.96	0.96	0.97	0.98	0.98	0.99	1.04	1.01	1.02
Ro	0.96	1.60	0.63	0.61	1.31	0.77	1.03	1.43	0.61	1.30	1.24	0.72	0.83	0.73	0.82	0.80	0.90	1.35	1.20	1.16
Ph	0.03	0.11	2.46	1.86	-2.78	1.28	0.30	-2.45	1.26	-2.87	-1.41	1.64	0.30	1.37	0.87	1.50	1.15	-1.27	-3.00	-0.78
Li	-0.56	-0.44	1.43	0.11	-1.13	2.15	-0.85	-1.38	0.02	-1.16	-1.69	1.02	3.00	0.08	-0.22	1.15	0.27	-1.50	-0.60	0.30

2.3. Sequence profiling

Sequence profiling is completed with the classical sliding window method. Briefly, a window slides from the N-terminal to C-terminal of the query protein sequence. The mean propensity value of the residues in the window is then assigned to the residue in the middle. At the N- and C- termini, we use asymmetric windows to avoid omitting prediction examples. Different window sizes of 1, 3, 5, 7, 9 and 11 are tested. If an index correlates negatively to the B cell epitope residues, it is then multiplied by -1 when used in prediction; this process makes the index have a positive predictive power.

2.4. Performance measures

Receiver operating characteristics (ROC) curves are constructed by varying the prediction threshold and plot the false-positive proportion (1-specificity) on the x-axis against the true positive proportion (sensitivity) on the y-axis [32]. The area under the ROC curve (*Aroc*) is used as the performance measure. For a random prediction, *Aroc* equals 0.5; for a perfect method, *Aroc* equals 1. Empirically, a prediction with an *Aroc* between 0.9 and 1 would be considered as "excellent"; 0.8-0.9, "good"; 0.7-0.8, "fair"; 0.6-0.7, "poor"; 0.5-0.6, "fail". More practically, an *Aroc* value higher than 0.7 indicates a useful prediction performance [33]. In this study, ROC curves and related performance measures are constructed, visualized and calculated with the ROCR package [34].

2.5. *Programming and Statistics*

All analyses are implemented in Perl or R; the latter is a language and environment for statistical computing and graphics. For predictions of special interest, they are further bootstrapped 1000 times. Random predictions are simulated through permuting the prediction results 1000 times. The *Aroc* differences of different indices are evaluated with *t*-tests.

3.   Results

3.1. *Performances of indices on different data sets*

For each index and for each data set, sliding window sizes of 1, 3, 5, 7, 9 and 11 were used for B cell epitope residue prediction. The performance of the indices on each data set is shown in Table 2. The results for the window size at which the majority of indices reach maximum performance are shown. For example, "Pellequer(7)" means the *Aroc* values are from testing the Pellequer dataset using a sliding window size of 7. As shown in Table 2, all indices performed poorly on the Pellequer, AntiJen, HIV, and DiscoTope data sets. However, their performance improved significantly when applied to the well-annotated HEL data set.

Table 2. Index performance on 5 data sets.

	Pellequer(7)	AntiJen(11)	HIV (11)	DiscoTope(9)	HEL (1)
Rk	0.627	0.561	0.588	0.591	0.794
Rc	0.637	0.564	0.586	0.608	0.752
Ro	0.609	0.566	0.575	0.583	0.787
Rb	0.633	0.567	0.589	0.610	0.772
Ph	0.655	0.565	0.586	0.622	0.733
Li	0.620	0.572	0.567	0.613	0.711

3.2. *ROC curves and statistical analysis*

According to Table 2, each index showed its best performance on the HEL data set at a sliding window size of 1. The prediction using the relative connectivity, Parker's hydrophilicity and Levitt's index were further bootstrapped 1000 times. Random predictions were simulated by permuting the prediction results 1000 times (see Table 3). Statistical tests showed that the relative connectivity index performed significantly better than the Parker's hydrophilicity and Levitt's index and the performance of all 3 were significantly better than random ($P < 2.2 \times 10^{-16}$). Their ROC curves were shown in Figure 1. On the HEL data set at a sliding window size of 1, the prediction accuracy, sensitivity and specificity of relative connectivity reached 73.6%, 73.0% and 75.0% respectively, with an area under the curve of 0.796.

Table 3. Comparing *Rk* performance with *Ph*, *Li* and Random on the HEL data set.

	Mean *Aroc* ± Standard Error
Rk	0.796 ± 0.001
Ph	0.733 ± 0.001
Li	0.708 ± 0.002
Random	0.497 ± 0.002

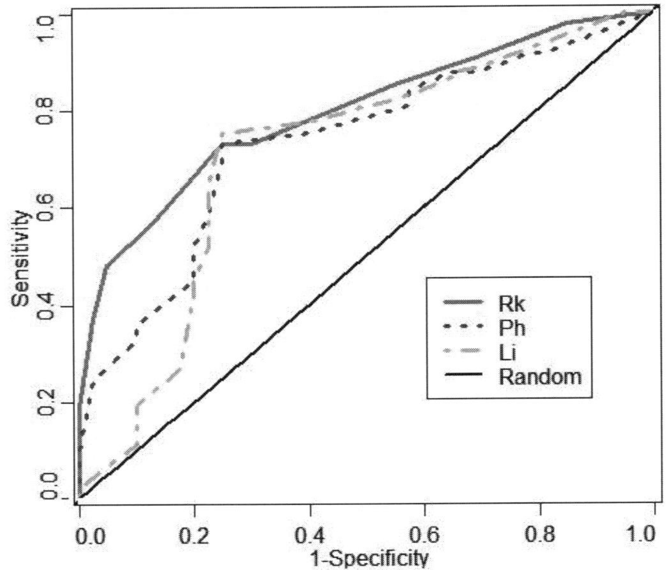

Figure 1: ROC curves for *Rk*, *Ph*, *Li* and Random. Based on the HEL data set at a window size of 1.

3.3. *Similar performance between surface and B cell epitope residue prediction*

The solvent accessible area of hen egg white lysozyme is computed from its PDB structure (1HEL) with the NACCESS program using default parameters. Surface residues are assigned based on the solvent accessible area at different cutoff values of 1, 10, 20, 50 and 100 Å^2. Prediction of surface residues was carried out using a sliding window size of 1. *Aroc* values are shown in Table 4, where "Bepi" means B cell epitope residue prediction and "Surf1" means surface residue prediction at the 1 Å^2 cutoff and so on. As shown in Table 4, the performance of B cell epitope residue prediction is very similar to that of protein surface residue prediction, especially at the 10 and 20 Å^2 cutoffs.

Table 4. Performance similarity between surface and B cell epitope residue prediction.

	Bepi	Surf1	**Surf10**	**Surf20**	Surf50	Surf100
Rk	**0.794**	0.758	**0.794**	**0.791**	0.778	0.731
Rc	**0.752**	0.727	**0.773**	**0.767**	0.732	0.594
Ro	**0.787**	0.753	**0.787**	**0.784**	0.766	0.707
Rb	**0.772**	0.744	**0.786**	**0.777**	0.744	0.624
Ph	**0.733**	0.713	**0.709**	**0.734**	0.673	0.574
Li	**0.711**	0.722	**0.711**	**0.729**	0.666	0.527

4. Discussions

4.1. *Relative connectivity can be a good choice for B cell epitope prediction*

The most common method for B cell epitope prediction is the sliding window and amino acid index based sequence profiling method [4-17]. However, its performance has been disputed [21-23]. In a recent report, 484 amino acid indices in the AAindex database were assessed with sequence profiling methods and the results showed that even the best set of amino acid indices performed only marginally better than random [24]. This indicates that better methods or new amino acid indices are needed for B cell epitope prediction. New methods such as neural networks, hidden Markov models and support vector machines have been applied to B cell epitope prediction very recently [25, 35-37]. However, the performance improvements are still limited.

In a previous study, we built four new amino acid indices based on the topological parameters of residue networks constructed from 640 representative PDB structures [26]. The Parker's hydrophilicity index and the Levitt's index have been confirmed to be the best two indices so far for B cell epitope prediction [12, 25]. Compared with the two indices, the 4 network topology based indices showed better performance in protein surface residue prediction [26]. Since surface accessibility implies antibody accessibility [27, 28], we wondered if the new indices would also show better performance in B cell epitope residue prediction.

Indeed, the results of this study show that the network based indices, especially relative connectivity, perform better than the Parker's hydrophilicity and Levitt's index on the well-annotated HEL data set. For other data sets, the performances of all indices are comparable. On the HEL data set at the sliding window size of 1, the relative connectivity performed significantly better than the Parker's hydrophilicity and Levitt's index and the performance of all 3 was significantly better than random prediction. The prediction accuracy, sensitivity and specificity of relative connectivity reached 73.6%, 73.0% and 75.0% respectively, with an area under the curve of 0.796. In fact, even on a poor performing data set (*e.g.* AntiJen data set), all the predictions were still significantly better than random (data not shown). We concluded that the network topology based indices, especially the relative connectivity, are useful indices for B cell epitope residue prediction.

4.2.   *Performance depends on the data sets*

A trend that better performance depends on better-annotated data set was observed. According to Table 2, all indices performed poorly on the Pellequer, AntiJen, HIV, and DiscoTope data sets. However, their performance improved significantly when applied to the HEL data set. The HEL data set is based on the well-studied model antigen hen egg white lysozyme. The protein was densely annotated with 19 epitopes, most of them derived from high quality structures of antigen-antibody complexes. The Discotope data set is also derived from crystals of antigen-antibody complexes. However, in this data set, each antigen sequence is only annotated with one epitope. Most data in the left 3 data sets is annotated with information from overlapping peptide experiments, which might have errors because a peptide can bind an antibody even if some residues of the peptide are not interacting with the antibody [29]. Even so, the trend in these 3 data sets is also obvious. The performance on the fully annotated Pellequer data set is better than the less annotated AntiJen and HIV data sets. One can also expect that some false positive predictions are actually undiscovered B cell epitope residues.

Thus, we concluded that the low performance of B cell epitope prediction is not only due to the methodology used, but the data set as well. The limited performance improvements observed with new methods might also be due to the data set itself. The importance of data set for B cell epitope prediction has also been addressed in a workshop very recently [38]. Another interesting phenomenon was also observed, that is the better the data set is annotated, the smaller the optimum sliding window size is (see Table 2 and data not shown).

4.3.   *The relationship between surface residues and B cell epitope prediction*

The problem of surface residue prediction is related to that of B cell epitope prediction, due to the requirement for epitopes to be surface accessible to interact with an antibody [27, 28]. In fact, most amino acid index based B cell epitope prediction methods, if not all, utilize this correlation. The Parker's hydrophilicty index [8] and β turn scale [15] are two good examples. In our previous study, the 4 network topology based amino acid indices showed a useful performance in surface residue prediction and they also correlated with hydrophobicity (or hydrophilicty) and β propensity [26].

In the current study, we found that the performance of B cell epitope residue prediction is very similar to that of protein surface residue prediction on the well-annotated HEL data set, especially at the at the 10 and 20 Å^2 cutoffs (see Table 4). It is proposed that any part of the accessible surface of a globular protein antigen can be recognized by antibodies, and the entire exposed surface represents a "continuum" of overlapping potential epitopes [39]. Therefore, we suggest that the performance in surface residue prediction might form a theoretical upper limit for the performance in B cell epitope residue prediction. As a B cell epitope is a context dependent immunological entity [38, 40, 41], a new paradigm for B cell epitope prediction may emerge, shifted from an "all B cell epitopes model" to a "single B cell epitope model", and from an "only antigen sequence based model" to a "multiple information based model." Besides the antigen sequence information, other information such as antigen structure [29], antibody sequence or mimotopes [42] are needed in the new generation of B cell epitope prediction methods.

Acknowledgments

We thank Dr Alex Gutteridge for copyediting the manuscript, giving valuable comments and helping us on R. This work was supported by grants from the Ministry of Education, Culture, Sports, Science and Technology and the Japan Science and Technology Agency. The computational resource was provided by the Bioinformatics Center, Institute for Chemical Research, Kyoto University. The support from NSFC project (30600138) is also gratefully acknowledged.

References

[1] Meloen, R.H., Puijk, W.C., Langeveld, J.P., Langedijk, J.P. and Timmerman, P., Design of synthetic peptides for diagnostics, *Curr Protein Pept Sci*, 4(4):253-260, 2003.

[2] Tanabe, S., Epitope peptides and immunotherapy, *Curr Protein Pept Sci*, 8(1):109-118, 2007.

[3] Naz, R.K. and Dabir, P., Peptide vaccines against cancer, infectious diseases, and conception, *Front Biosci*, 12:1833-1844, 2007.

[4] Kazim, A.L. and Atassi, M.Z., Prediction and conformation by synthesis of two antigenic sites in human haemoglobin by extrapolation from the known antigenic structure of sperm-whale myoglobin, *Biochem J*, 167(1):275-278, 1977.

[5] Hopp, T.P. and Woods, K.R., Prediction of protein antigenic determinants from amino acid sequences, *Proc Natl Acad Sci U S A*, 78(6):3824-3828, 1981.

[6] Hopp, T.P. and Woods, K.R., A computer program for predicting protein antigenic determinants, *Mol Immunol*, 20(4):483-489, 1983.

[7] Welling, G.W., Weijer, W.J., van der Zee, R. and Welling-Wester, S., Prediction of sequential antigenic regions in proteins, *FEBS Lett*, 188(2):215-218, 1985.

[8] Parker, J.M., Guo, D. and Hodges, R.S., New hydrophilicity scale derived from high-performance liquid chromatography peptide retention data: correlation of predicted surface residues with antigenicity and X-ray-derived accessible sites, *Biochemistry*, 25(19):5425-5432, 1986.

[9] Jameson, B.A. and Wolf, H., The antigenic index: a novel algorithm for predicting antigenic determinants, *Comput Appl Biosci*, 4(1):181-186, 1988.

[10] Hopp, T.P., Use of hydrophilicity plotting procedures to identify protein antigenic segments and other interaction sites, *Methods Enzymol*, 178:571-585, 1989.

[11] Menendez-Arias, L. and Rodriguez, R., A BASIC microcomputer program for prediction of B and T cell epitopes in proteins, *Comput Appl Biosci*, 6(2):101-105, 1990.

[12] Pellequer, J.L., Westhof, E. and Van Regenmortel, M.H., Predicting location of continuous epitopes in proteins from their primary structures, *Methods Enzymol*, 203:176-201, 1991.

[13] Maksyutov, A.Z. and Zagrebelnaya, E.S., ADEPT: a computer program for prediction of protein antigenic determinants, *Comput Appl Biosci*, 9(3):291-297, 1993.

[14] Pellequer, J.L. and Westhof, E., PREDITOP: a program for antigenicity prediction, *J Mol Graph*, 11(3):204-210, 191-202, 1993.

[15] Pellequer, J.L., Westhof, E. and Van Regenmortel, M.H., Correlation between the location of antigenic sites and the prediction of turns in proteins, *Immunol Lett*, 36(1):83-99, 1993.

[16] Alix, A.J., Predictive estimation of protein linear epitopes by using the program PEOPLE, *Vaccine*, 18(3-4):311-314, 1999.

[17] Odorico, M. and Pellequer, J.L., BEPITOPE: predicting the location of continuous epitopes and patterns in proteins, *J Mol Recognit*, 16(1):20-22, 2003.

[18] Nakai, K., Kidera, A. and Kanehisa, M., Cluster analysis of amino acid indices for prediction of protein structure and function, *Protein Eng*, 2(2):93-100, 1988.

[19] Tomii, K. and Kanehisa, M., Analysis of amino acid indices and mutation matrices for sequence comparison and structure prediction of proteins, *Protein Eng*, 9(1):27-36, 1996.

[20] Kawashima, S. and Kanehisa, M., AAindex: amino acid index database, *Nucleic Acids Res*, 28(1):374, 2000.

[21] Hopp, T.P., Retrospective: 12 years of antigenic determinant predictions, and more, *Pept Res*, 6(4):183-190, 1993.

[22] Van Regenmortel, M.H. and Pellequer, J.L., Predicting antigenic determinants in proteins: looking for unidimensional solutions to a three-dimensional problem?, *Pept Res*, 7(4):224-228, 1994.

[23] Hopp, T.P., Different views of protein antigenicity, *Pept Res*, 7(4):229-231, 1994.

[24] Blythe, M.J. and Flower, D.R., Benchmarking B cell epitope prediction: underperformance of existing methods, *Protein Sci*, 14(1):246-248, 2005.

[25] Larsen, J.E., Lund, O. and Nielsen, M., Improved method for predicting linear B-cell epitopes, *Immunome Res*, 2(2, 2006.

[26] Huang, J., Kawashima, S. and Kanehisa, M., New amino acid indices based on residue network topology, *Genome Informatics*, 18(1):in press, 2007.

[27] Novotny, J., Handschumacher, M., Haber, E., Bruccoleri, R.E., Carlson, W.B., Fanning, D.W., Smith, J.A. and Rose, G.D., Antigenic determinants in proteins coincide with surface regions accessible to large probes (antibody domains), *Proc Natl Acad Sci U S A*, 83(2):226-230, 1986.

[28] Thornton, J.M., Edwards, M.S., Taylor, W.R. and Barlow, D.J., Location of 'continuous' antigenic determinants in the protruding regions of proteins, *Embo J*, 5(2):409-413, 1986.

[29] Haste Andersen, P., Nielsen, M. and Lund, O., Prediction of residues in discontinuous B-cell epitopes using protein 3D structures, *Protein Sci*, 15(11):2558-2567, 2006.

[30] Huang, J. and Honda, W., CED: a conformational epitope database, *BMC Immunol*, 7:7, 2006.

[31] Hubbard, S.J. and Thornton, J.M., NACCESS, *Department of Biochemistry and Molecular Biology, University College London,* 1993.

[32] Swets, J.A., Measuring the accuracy of diagnostic systems, *Science*, 240(4857):1285-1293, 1988.

[33] Lund, O., Nielsen, M., Lundegaard, C., Kesmir, C. and Brunak, S., Immunological Bioinformatics, *The MIT Press*, 2005.

[34] Sing, T., Sander, O., Beerenwinkel, N. and Lengauer, T., ROCR: visualizing classifier performance in R, *Bioinformatics*, 21(20):3940-3941, 2005.

[35] Saha, S. and Raghava, G.P., Prediction of continuous B-cell epitopes in an antigen using recurrent neural network, *Proteins*, 65(1):40-48, 2006.

[36] Sollner, J. and Mayer, B., Machine learning approaches for prediction of linear B-cell epitopes on proteins, *J Mol Recognit*, 19(3):200-208, 2006.

[37] Chen, J., Liu, H., Yang, J. and Chou, K.C., Prediction of linear B-cell epitopes using amino acid pair antigenicity scale, *Amino Acids*, 2007.

[38] Greenbaum, J.A., Andersen, P.H., Blythe, M., Bui, H.H., Cachau, R.E., Crowe, J., Davies, M., Kolaskar, A.S., Lund, O., Morrison, S., Mumey, B., Ofran, Y., Pellequer, J.L., Pinilla, C., Ponomarenko, J.V., Raghava, G.P., van Regenmortel, M.H., Roggen, E.L., Sette, A., Schlessinger, A., Sollner, J., Zand, M. and Peters, B., Towards a consensus on datasets and evaluation metrics for developing B-cell epitope prediction tools, *J Mol Recognit*, 20(2):75-82, 2007.

[39] Benjamin, D.C., Berzofsky, J.A., East, I.J., Gurd, F.R., Hannum, C., Leach, S.J., Margoliash, E., Michael, J.G., Miller, A., Prager, E.M. and et al., The antigenic structure of proteins: a reappraisal, *Annu Rev Immunol*, 2:67-101, 1984.

[40] Van Regenmortel, M.H.V., Mapping Epitope Structure and Activity: From One-Dimensional Prediction to Four-Dimensional Description of Antigenic Specificity, *Methods*, 9(3):465-472, 1996.

[41] Van Regenmortel, M.H., Immunoinformatics may lead to a reappraisal of the nature of B cell epitopes and of the feasibility of synthetic peptide vaccines, *J Mol Recognit*, 19(3):183-187, 2006.

[42] Huang, J., Gutteridge, A., Honda, W. and Kanehisa, M., MIMOX: a web tool for phage display based epitope mapping, *BMC Bioinformatics*, 7:451, 2006.

COMPARATIVE PAIR-WISE DOMAIN-COMBINATIONS FOR SCREENING THE CLADE SPECIFIC DOMAIN-ARCHITECTURES IN METAZOAN GENOMES

SHUICHI KAWASHIMA[1]
shuichi@hgc.jp

TAKESHI KAWASHIMA[2,3]
kawashima38@gmail.com

NICHOLAS H PUTNAM[2]
nhputnam@lbl.gov

DANIEL S ROKHSAR[2]
DSRokhsar@lbl.gov

HIROSHI WADA[4]
hwada@biol.tsukuba.ac.jp

MINORU KANEHISA[1,5]
kanehisa@kuicr.kyoto-u.ac.jp

[1] *Human Genome Center, Institute of Medical Science, University of Tokyo, 4-6-1 Shirokanedai, Minato-ku, Tokyo 108-8639, Japan*
[2] *Department of Molecular and Cell Biology, University of California, Berkeley, LSA Bldg. #3200, Berkeley, CA 94720-3200, USA*
[3] *Japan Sociery for the Promotion of Science, 8 Ichibancho, Chiyoda-ku, Tokyo 102-8472, Japan*
[4] *Graduate School of Life and Environmental Sciences, University of Tsukuba, 1-1-1 Tennoudai, Tsukuba, Ibaraki 305-8572, Japan*
[5] *Bioinformatics Center, Institute for Chemical Research, Kyoto University, Gokasho, Uji, Kyoto 611-0011, Japan*

In the evolution of the eukaryotic genome, exon or domain shuffling has produced a variety of proteins. On the assumption that each fusion event between two independent protein-domains occurred only once in the evolution of metazoans, we can roughly estimate when the fusion events were happened. For this purpose, we made phylogenetic profiles of pair-wise domain-combinations of metazoans. The phylogenetic profiles can be expected to reflect the protein evolution of metazoan. Interestingly, the phylogenetic tree of metazoans, derived from the profiles, supported the "Ecdysozoa hypothesis" that is one of the major hypotheses for metazoan evolution. Further, the phylogenetic profiles showed the candidates of genes that were required for each clade-specific features in metazoan evolution. We propose that comparative proteome analysis focusing on pair-wise domain-combinations is a useful strategy for researching the metazoan evolution. Additionally, we found that the extant ecdysozoans share only fourteen domain-combinations in our profiles. Such a small number of ecdysozoan-specific domain-combinations is consistent with the extensive gene-losses through the evolution of ecdysozoans.

Keywords: protein evolution; domain-combination; ecdysozoan; coelomata; phylogenetic tree; metazoa.

1. Introduction

In the evolution of the eukaryotic genomes, exon shuffling or domain shuffling has produced a variety of proteins. Fusion- and fission-events of domains or exons yield new domain-architectures of proteins that may possess novel functions. However, the linkage between such a gain of new domain-architectures and the evolution of metazoans (or animals) remains to be elucidated.

Cloning of the cadherin genes from amphioxus and the finding of its unique domain-architectures provide an opportunity to reconsider the conventional idea of deuterostome phylogeny [15, 16]. Oda and his colleagues found that the domain-architectures of the amphioxus cadherins are similar to invertebrate cadherins rather than tunicate and vertebrate cadherins, although cephalochordata (amphioxus) was long considered as the closest sister group to vertebrates. More recently, two groups reported the close relationship between vertebrate and tunicate by the large amount of sequences from key taxa of metazoans [3, 5].

Considering the case of cadherin, it is expected that comparative analysis of domain-architectures can reveal the evolutional history. Actually, Patthy found that several chordate-specific domain-architectures are related to vertebrate or chordate-specific features [18]. However, it is difficult to compare with the details of domain-architectures for multiple proteomes. In order to address the rapid increase of the draft genomes from variety of metazoan taxa, a convenient method to detect genes including clade-specific domain-architectures is required. Here we propose a new method for comparative domain-architectures, which is useful for comparing among multiple proteomes. Previously, we made the lists of genes with clade-specific domain-architectures [12]. It is noteworthy that the list includes a lot of known genes that have been annotated as their chordate- and vertebrate-specific domain-architectures. In the method, not single domains but pair-wise domain-combinations are regarded as a unit of phylogenetic analysis. The phylogenetic profiles of pair-wise domain-combinations allowed us to survey the timing of the gain and loss of domain-combinations, or gene-fusion and -fission events in the metazoan phylogeny. In the analysis, we found that several vertebrate-specific domain-combinations seem to be required for gaining the vertebrate-specific features such as "cartilage", "auditory systems" or "tight junction" [12].

Here we extend the analysis of the evolution of the domain architectures to fifteen metazoan genomes and three other higher eukaryotes (fungi, slime mold and plant). Fig. 1 shows the schematic view of the major hypotheses for the phylogeny

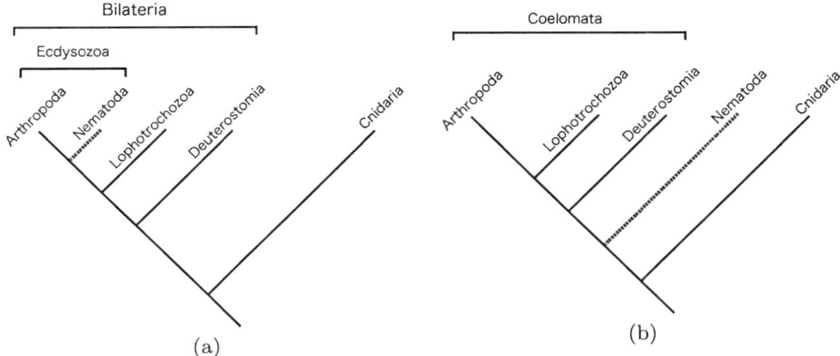

Fig. 1.   (a) Ecdysozoa hypothesis. (b) Coelomata hypothesis.

of the metazoa. We highlight particularly the topology of the four branches of the tree, that is Deuterostomia, Arthropoda, Nematoda and Non-bilaterians (including Cnidaria), respectively.

In Fig. 1(a), the nematoda is related to arthropoda in a clade of molting protostomes, termed "Ecdysozoa". Both the Ecdysozoa and Deuterostomia are members of a clade termed "Bilateria" which also includes the third major group termed "Lophotrochozoa" which we do not discuss here. Other non-bilateria groups including Cnidaria are placed at the out-group position of Bilateria. Fig. 1(b) differs from the Fig. 1(a) only in the branch of Nematoda. In Fig. 1(b), Nematoda is sister to the "Coelomata" which includes both Deuterostomia and Arthropoda. Because of their complicated protein domain-architectures, the relationships between protein evolution and its functional innovation is not a simple question, especially in metazoan evolution. Even if there are limited numbers of protein domains in metazoans, their shuffling, fusions and combinations can make a variety of proteins. For example, in the context of two genomes from different species coded the same four protein domains (A, B, C and D), if one species has two genes constructed from "A and B" and "C and D", and meanwhile another species have genes construed from "A and D" and "B and C", can we say that whether they have the same two genes or not? Mutual best-hit search between two species is a simple and convenient method for comparative proteomes, but it may not be powerful as shown in Fig. 2. In Fig 2, a pair of proteins are related by the mutual best-hit. From the view point of domain-combination, however, these are not treated as related genes because no domain-fusion event was shared between these two proteins. Thus, the timing of the domain-fusion event will be revealed by the phylogenetic analysis of domain-architectures. We found there are large differences among the number of clade-specific domain-combinations.

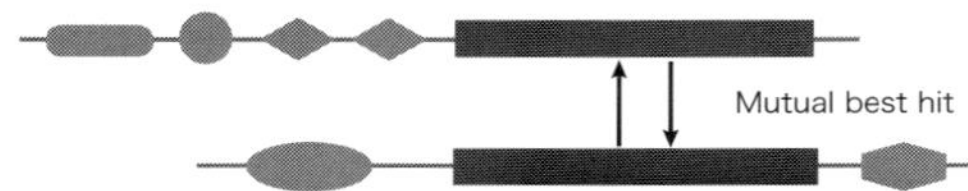

Fig. 2. Schematic view of mutual best hit between two genes containing different domain-architectures.

We can reconstruct a phylogenetic tree from our phylogenetic profiles of pair-wise domain-combinations, when we regard these profiles as a discrete data matrix. We produced such a phylogenetic tree with Phylip Program Package [7] and TreeView [17].

In this phylogenetic tree, two insects and two nematodes are grouped into the same clade. This supports the "Ecdysozoa hypothesis" rather than the "Coelomata hypothesis". Additionally, we found that the very small number of domain-combinations seemed to be Ecdysozoa-specific.

2. Materials and Methods

2.1. *Data*

The following 18 proteomes are used for comparative analysis. Human (*Homo sapiens*): Ensembl build 41 from Ensembl [24], [10], Mouse (*Mus musculus*): Ensembl build 41, Norway Rat (*Rattus norvegicus*): Ensembl build 41, Chicken (*Gallus gallus*): Ensembl build 41, Flog (*Xenopus tropicalis*): JGI v4.1 from JGI [27], Zebra fish (*Danio rerio*): Ensembl build 41, Fugu (*Takifugu rubripes*): JGI version 4.0 from JGI, Ascidian (*Ciona intestinalis*): JGI version 2.0, Amphioxus (*Branchiostoma floridae*): JGI version 1.0, Sea Urchin (*Strongylocentrotus purpuratus*): NCBI gene build 2 version 1 on Baylor's assembly Spur_v2.1 from www.ncbi.nlm.gov, Fly (*Drosophila melanogaster*): Ensembl build 41, Mosquito (*Anopheles gambiae*): Ensembl build 41, Nematodes (*Caenorhabditis elegans*): Wormbase release WS164 from Wormbase [30] and (*Caenorhabditis briggsae*): Wormbase release WS165, Sea anemone (*Nematostella vectensis*): JGI Annotation of Nematostella genome version 1, Fungi (*Saccharomyces cerevisiae*): genome-ftp.stanford.edu saccharomyces_cerevisiae.gff created on july7-2004, Social amoeba (*Dictyostelium discoideum*): dictyBase [23] Full Chromosomes made 10/05/2004 (primary features made 7/11/2005), and Plant (*Arabidopsis thaliana*): TAIR6 updated 11.2005 from NCBI [28].

2.2. *Construction of the pair-wise domain-combinations*

Pfam database (Pfam_ls, release 16.0) was used as the reference domain database [8]. Hammer search for all of the proteomes against Pfam was executed under a threshold set as $1.0e^{-3}$ [26]. Outputs of hmmpfam are parsed by scripts written in Ruby programming language [29] with BioRuby library [22].

The definition of the pair-wise domain-combination is as follows. As a fundamental idea, we enumerate unique pair-wise domain-combinations without taking into account of the order from N-terminus to C-terminus of proteins. For example, when four domains "A", "B", "A" and "C" are lined on a gene in this order, three pair-wise domain-combinations "A and B", "B and C" and "A and C" can be counted. As shown above example, we do not consider the direction of domains on the gene. Therefore, we treat "A and B" and "B and A" as the same combinations. Further, we do not regard the domain repetition "A and A" as pair-wise domain-combinations.

2.3. *Construction of the phylogenetic profiles*

The phylogenetic profiles of domain-combinations of the 18 species are made as follows. For each domain-combination, we count the number of genes having the both Pfam domains which belong to the domain-combination for each organism and put the numbers in the order of the 18 organisms as shown below.

For example, if the numbers of genes coded a domain-combination are 3, 2, 1,

0, 0, 0 in species S_1, S_2, S_3, S_4, S_5, S_6, respectively, the profile is treated as the following:

$$[S_1, S_2, S_3, S_4, S_5, S_6] \mapsto [3, 2, 1, 0, 0, 0]$$

The whole phylogenetic profiles are shown in the supplemental Table S1. In the table, each row represents a phylogenetic profile of a pair-wise Pfam domain-combination. The first and the second column represents two accession numbers of the Pfam domains which belong to the combination, respectively. Each phylogenetic profile is shown as a set of numbers written in columns from the third to the 20th. From the third to the 20th column, the corresponding 18 species are, *H.sapience*, *M.musculus*, *R.norvegicus*, *G.gallus*, *X.tropicalis*, *D.rerio*, *T.rubripes*, *B.floridae*, *C.intestinalis*, *S.purpuratus*, *D.melanogaster*, *A.gambiae*, *C.elegans*, *C.briggsae*, *N.vectensis*, *D.discoideum*, *S.cerevisiae* and *A.thaliana*, respectively.

Phylogenetic profiles of domain-combinations for the super phyla are made as follows. The columns of phylogenetic profiles of the 18 species (in Table S1) are divided into the following three super phyla or an outgroup. Ten species are assigned to Deuterostomia (*H.sapience*, *M.musculus*, *R.norvegicus*, *G.gallus*, *X.tropicalis*, *D.rerio*, *T.rubripes*, *B.floridae*, *C.intestinalis*, *S.purpuratus*), two species to Arthropoda (*D.melanogaster*, *A.gambiae*), two species to Nematoda (*C.elegans*, *C.briggsae*), four species to Out-group (*N.vectensis*, *S.cerevisiae*, *D.discoideum*, *A.thaliana*). The result of phylogenetic profiles of the super phyla is shown in Table 1.

Phylip program [7] was used for making the phylogenetic tree from the phylogenetic profile of pair-wise domain-combinations. We converted the Table S1 into binary data (i.e. 0 to 0 and otherwise to 1) and used it as input data for Pars that is a parsimony program included in the Phylip package. The phylogenetic tree was shown in Fig. 3.

3. Results

In our previous study, we made lists of the several clade-specific genes in deuterostomes. The functional annotations and experiments of the genes strongly suggested links between their functions and their clade-specific features; for example, chordate-specific genes included notochord related genes and vertebrate-specific genes included cartilage specific genes, etc. [12].

These results mean that our method is effective for understanding the relationships between protein evolution and biological features in the evolution of deuterostomes. Here, we would like to focus on another key animal clade, Ecdysozoa, which includes two big sub-clades, Nematoda and Arthropoda. Well-annotated genomes of *C.elegans* and *D.melanogaster* are available as representatives of both clades. We selected the genes containing the ecdysozoan-specific domain-combinations. Interestingly, we found only fourteen genes satisfying the above condition and seven of the fourteen were common to the ecdysozoans (Table 1 "_AN_" and Table 2).

Table 1. The numbers of domain-combinations shared among four super phyla.

The type of profiles[a]	The number of shared domain-combinations[b]	The number of shared class-I domain-combination[c]	The number of shared class-II domain-combination[d]
D A N O	1,468	0	1,468
D A N _	298	22	276
D A _ O	372	0	372
D _ N O	120	0	120
_ A N O	0	0	0
D A _ _	236	21	215
D _ N _	106	4	102
_ A N _	18	4	14
D _ _ O	572	0	572
_ A _ O	9	0	9
_ _ N O	10	0	10
D _ _ _	3,702	236	3,466
_ A _ _	100	12	88
_ _ N _	176	17	159
_ _ _ O	1,297	0	1,297
Total	8,484	316	8,168

Note: [a] D: Deuterostomia (*H.sapience*, *M.musculus*, *R.norvegicus*, *G.gallus*, *D.rerio*, *T.rubripes*, *B.floridae*, *C.intestinalis* and *S.purpuratus*). A: Arthropoda (*D.melanogaster* and *A.gambiae*). N: Nematoda (*C.elegans* and *C.briggsae*). O: Out-group of bilaterians (Cnidaria (*N.vectensis*), and other higher eukaryotes (*S.cerevisiae*, *D.discoidium* and *A.thaliana*)). Underscore ("_") means that the corresponding organismal-group do not have the domain-combinations. (E.g., the row of "DAN_" represents the number of domain-combinations that are shared between Deuterostomia, Arthropoda and Nematoda but not with out-group of bilaterians). [b] The total of the third column and fourth column is shown in the second column. [c] The class-I domain-combination defined as the each individual domains is not found in Out-group. [d] The class-II domain-combination defined as the both of the individual domains are found in Out-group.

Thus, the number of domain-combinations gained in the ecdysozoans is quite small, compared to that gained in deuterostomia (3466 in Table 1, "D _ _ _"). We examined the meaning of differences between these two numbers blow.

We constructed a phylogenetic tree from the phylogenetic profiles. (Fig. 3). In the tree, the two species of Arthropoda (*D.melaogaster* and *A.gambiae*) and the two species of Nematoda (*C.elegans* and *C.briggsae*) form a clade. This means that information from gain and loss of domain-combinations supports the "Ecdysozoa hypothesis" rather than the "Coelomata hypothesis".

Table 1 shows the phylogenetic profile of the domain-combinations from three clades of super phyla (Arthropoda, Nematoda and Deuterostomia) and the out-groups (cnidarian, fungi and plant). Since all combinations in the fourth column of Table 1 are of two modules both of which are found in out-group species, it is expected that these numbers represent the gains of combinations from pre-existing domains. Because all of the single domains counted on the fourth column in Table 1 are found in out-group species, it is expected that these numbers represent the gains of

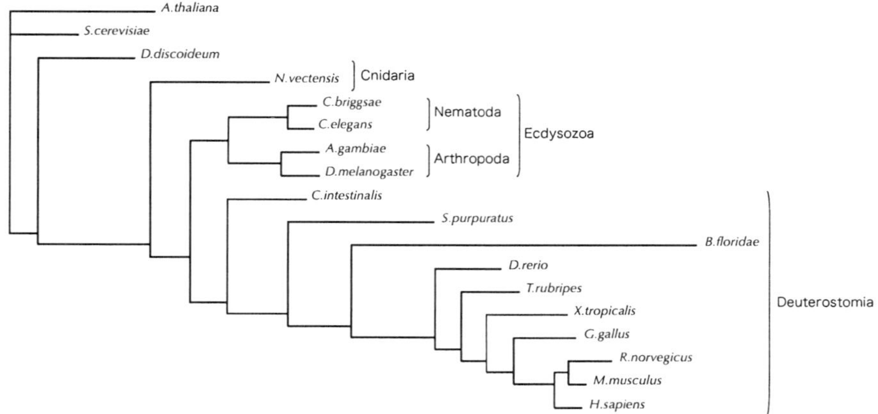

Fig. 3.   Parsimonious tree based on domain-combinations from 18 species.

domain-combinations from pre-existing two individual domains. In the Table 1, the number of common domain-combinations within Deuterostomia and Arthropoda is 215. In contrast, the number of common domain-combinations within Arthropoda and Nematoda is only 14. Thus, if we ascribed the gain of domain-combinations to the important factors which influence the topology of the phylogenetic tree, these two numbers may support the "Coelomata hypothesis".

However, the counterpart that is the loss of domain-combinations also influences the topology. Therefore the topology of branches supports the "Ecdysozoa hypothesis". The number of loss of domain-combinations which is common to Arthropoda and Nematoda is 572 from out-group, and the corresponding number to Deuterostomia and Nematoda is merely 10. This result supports a scenario in which extensive gene losses have occurred in the evolution of Ecdysozoans. This data is consistent with the result from comparative proteomes between human and cnidarians [13, 14, 19, 20]. EST analyses of Cnidaria that is one of the animal clades as out-group of bilaterians (Fig. 1) revealed that a substantial number of the ESTs appeared to be vertebrate homologs but no counterpart in the fly or worm.

Interestingly, Nematostella is located on the basal position of metazoans in Fig. 3, even though the number of mutual best-hit genes between human and nematostella is larger than those between others. This result suggests an advantage of our method which can distinguish the two sequences of similar proteins that have different domain-architectures. The best-hit search between human and Nematostella may possibly include the best-hit single domains which are partly constituting more complicated domain-architectures (Fig. 2). In fact, 3,271 domain-combinations are found only in Deuterostomia, although both of the single domains for the all of domain-combinations are found in an out-group (column 4 in Table 2). The 3,271 domain-combinations may be critical to the evolution of Deuterostomes.

Table 2. The list of the domain-combinations that are ecdysozoan-specific and that are shared between Nematodes (*C.elegans* ans *C.briggsae*) and Arthropods (*D.melanogaster* and *A.gambiae*). The Combinations of the two Pfam-domains ("Pfam accession" of the first column and second column) are found at least one gene in both of Nematodes and Arthropods.

Pfam accession	Pfam accession	Gene name[a]	Gene annotation[a]
PF06421	PF08477	Waw (fly)	Elongation factor-type GTP-binding protein
PF01011	PF07714	PEK (fly)	Unknown
PF00024	PF00100	NompA (fly)	Detection of mechanical stimulus during sensory perception of sound
PF00989	PF07885	Egl-2 (worm)	Voltage-gated potassium channel
PF00520	PF00989	Egl-2 (worm)	Voltage-gated potassium channel
PF00047	PF06582	Ketn-1 (worm)	Actin-bingind protein
PF00014	PF02014	Spon-1 (worm)	Extracellular matrix proteins
PF00014	PF06468	Spon-1 (worm)	Extracellular matrix proteins
PF00795	PF01230	Nft-1 (worm)	Carbon-nitrogen hydrolase
PF01822	PF02485	Sqv-6 (worm)	Peptide O-xylosyltransferase
PF00023	PF05186	F34D10.7 (worm)	Unknown
PF00096	PF01426	MTA1-like (fly)	Interacts with two or more components of the EGF/RAS signalling pathway
PF00041	PF01682	CG17839 (fly)	Unknown
PF00047	PF01682	CG17839 (fly)	Unknown

Note: [a] A representative gene name is shown in the third column ("Gene name"). Each annotation for the gene is shown in the fourth column ("Gene annotation"). Gene names and annotations are quoted from the wormbase [30] and the flybase [25].

4. Discussion

In this research, we employed the pair-wise domain-combinations rather than combination of more larger number of domains. e.g. ternary or quartette. Because we have focused evolutionary events when two different domains fused into one gene, not domain-architectures on genes itself. For another reason, the number of the data will be reduced if we considered more large number of domain combination. In addition, it is not suitable to utilize the detailed domain-architectures of genes in draft genomes for phylogenetic analysis. Meanwhile, even if the gene-models are still draft, it would be rarely seen that the same false-positive domain-combinations are detected throughout all genomes from one clade. Thus, we suggest that pair-wise domain-combinations are useful unit for the comparative analysis of domain-architectures of "draft" invertebrate genomes.

The phylogenetic profile of the pair-wise domain-combinations allows us to reconstruct a phylogenetic tree by a parsimonious method. In the tree, the phylogenetic location of the most of the species are consistent with that of the "Ecdysozoa hypothesis" [1] except *C.intestinalis* (Fig. 3). Here, we focused on the two important points in the hypothesis. One is that the several groups of molting animals, like nematodes, insects and other arthropods, form one clade named as Ecdysozoa. Another, the Cnidarians are the closest animal group of Bilateria (Fig. 1(a)). For the past 10 years, several analyses for the phylogenetic tree of metazoans supported the

"Ecdysozoa" hypothesis but others supported another traditional scenario, "Coelomata hypothesis".

Wang and Caetano-Anolles reported a global phylogeny of 185 organisms, based on the phylogenetic profiles of the domain-architectures among their proteomes [21]. In this respected work, they made profiles from not pair-wise but detail classification of domain-architectures. Interestingly the sub-tree of their phylogeny corresponding to metazoans supports the "Coelomata hypothesis", in contrast with our phylogenetic tree which supports the "Ecdysozoa hypothesis" (Fig. 1(a) and Fig. 3). Our study is the first to support the "Ecdysozoa hypothesis" from the perspective of domain architectures.

Several recent genome analyses of Cnidarians revealed that a significant proportion of genes that are not in ecdysozoa are shared between cnidarians and vertebrates [13, 14, 19, 20]. This suggests that a lot of genes are lost from a variety of phyla in Bilateria after the divergence from the common ancestor of cnidarians. If we stand with the "Ecdysozoa hypothesis", the large (572) number of common-loss of domain-combinations in Insecta and Nematoda is consistent with their monophyletic evolution. The small number (14) of common-gain of domain-combinations suggests that the divergence of Insecta and Nematoda occurred at an early stage after the event when the losses of domain-combinations happened in the common ancestor of ecdysozoan.

Conceivably, large number of gene losses would be independently occurred in the both Arthropoda and Nematoda. Also in this case, loss of genes common in the both species can be secondarily detected. For examining this possibility, we are now analyzing ESTs of another Arthropod to survey their domain-combinations. It would be interesting that what-like domain-combinations can be found in Arthropod other than Insecta.

A tunicate, *C. intestinalis*, locates on the basal position of Deuterostomia in both of the trees in our data as well as in data by Wang and Caetano-Anolles [21]. It seems to be misplacement because the echinoderms have been believed to be branched off earlier than Ciona or Amphioxus [9]. This can be explained by the substantial gene-loss in Tunicate [11]. The Tunicate-specific gene-loss didn't disrupt the position of Deuterostomia in the tree, but it displaces the location of *C.intestinalis* onto the basal position in Deuterostomia.

Thus, we think that the phylogenetic trees made from domain-architectures are not so robust as molecular phylogenetic trees. However, we would like to emphasize the value of the phylogenetic profiles of domain-architectures. We can perceive the past evolutional history from the profiles. The genes common in each phylogenetic node suggest that they were essential genes for evolution of each clade. In Table 2, we listed the ecdysozoan-specific domain-combination genes.

In a recent study, Bashton and Chothia asked how domains affect each other in multi-domain proteins, finding that often they interact in ways that create possibly-associated but new functional attributes [2]. They surveyed the SCOP database and assembled 45 sets of proteins, each containing a multi-domain protein and homologous 1-domain proteins. In some cases, the combination and interaction of domains may changes the function of a single domain to a related but different one in the context of the more complicated protein. The strategy of our research is similar to the report by Bashton and Chothia. The novelty of our study is that we mapped the domain-combinations on the metazoan-phylogeny. This mapping-step gives us the additional information regarding metazoan evolutions. In the previous study, we proposed a possibility in which the genes containing vertebrate-specific domain-combinations cause several vertebrate innovations, like cartilage formation, tight junction and auditory system [12]. In this report, we found that only fourteen domain-combinations are shared in the four of the extant ecdysozoans (Table 1, column 4, Table 2). Because such global gene-losses seem to be happened in the ancestors of ecdysozoans, the remained fourteen genes are good candidates to study the evolution of the ecdysozoans.

Acknowledgments

We thank Dr. David Pile, Dr. David Hendrix and Dr. Keisuke Kawashima for critical reading of our manuscript. This work was supported by grants from the Ministry of Education, Culture, Sports, Science and Technology, the Japan Society for the Promotion of Science, and the Japan Science and Technology Corporation. The computational resource was provided by the Bioinformatics Center, Institute for Chemical Research, Kyoto University and the Super Computer System, Human Genome Center, Institute of Medical Science, University of Tokyo. This work was also supported by JSPS Postdoctral Fellowships for Research Abroad.

References

[1] Aguinaldo, A.M., Turbeville, J.M., Linford, L.S., Rivera, M.C., Garey, J.R., Raff RA, and Lake, J.A., Evidence for a clade of nematodes, arthropods and other moulting animals, *Nature*, 387:489–493, 1997.

[2] Bashton, M., and Chothia, C., The generation of new protein functions by the combination of domains., *Structure*, 15:85-99, 2007.

[3] Bourlat, S.J., Juliusdottir, T., Lowe, C.J., Freeman, R., Aronowicz, J., Kirschner, M., Lander, E.S., Thorndyke, M., Nakano, H., Kohn, A.B., Heyland, A., Moroz, L.L., Copley, R.R. and Telford, M.J., Deuterostome phylogeny reveals monophyletic chordates and the new phylum Xenoturbellida, *Nature*, 444:85–88, 2006.

[4] Cordaux, R., Udit, S., Batzer, M.A., and Feschotte, C., Birth of a chimeric primate gene by capture of the transposase gene from a mobile element, *Proc. Natl. Acad. Sci. U.S.A.*, 103:8101-8106, 2006.

[5] Delsuc, F., Brinkmann, H., Chourrout D., and Philippe, H., Tunicates and not cephalochordates are the closest living relatives of vertebrates, *Nature*, 439:965-968, 2006.

[6] Dopazo, H., and Dopazo, J., Genome-scale evidence of the nematode-arthropod clade, *Genome Biol.*, 6:R41, 2005.

[7] Felsenstein, J., PHYLIP (Phylogeny Inference Package) version 3.6. Distributed by the author. Department of Genome Sciences, University of Washington, Seattle, 2005.

[8] Finn, R.D., Mistry, J., Schuster-Bockler, B., Griffiths-Jones, S., Hollich, V., Lassmann, T., Moxon, S., Marshall, M., Khanna, A., Durbin, R., Eddy, S.R., Sonnhammer, E.L., and Bateman, A., Pfam: clans, web tools and services, *Nucleic Acids Res.*, 34:D247-51, 2006.

[9] Holland, L.Z., A chordate with a difference, *Nature*, 447:153–155, 2007.

[10] Hubbard, T.J., Aken, B.L., Beal, K., Ballester, B., Caccamo, M., Chen, Y., Clarke, L., Coates, G., Cunningham, F., Cutts, T., Down, T., Dyer, S.C., et al., Ensembl 2007, *Nucleic Acids Res.*, 35:D610-617, 2006.

[11] Hughes, A.L., and Friedman, R., Loss of ancestral genes in the genomic evolution of Ciona intestinalis, *Evol. Dev.*, 7:196-200, 2005.

[12] Kawashima, T., Kawashima, S., Tanaka, C., Murai, M., Yoneda, M., Putnum, N.H., Rokhser, D.S., Kanehisa, M., Satoh, N., and Wada, H., Module shuffling preformed essential role during the evolution of vertebrates, In preparation.

[13] Kortschak, R.D., Samuel, G., Saint, R., and Miller, D.J., EST analysis of the cnidarian Acropora millepora reveals extensive gene loss and rapid sequence divergence in the model invertebrates, *Curr Biol.*, 13:2190-2195, 2003.

[14] Miller, D.J., Ball, E.E., and Technau, U., Cnidarians and ancestral genetic complexity in the animal kingdom, *Trends Genet.*, 21:536-539, 2005

[15] Oda, H., Wada, H., Tagawa, K., Akiyama-Oda, Y., Satoh, N., Humphreys, T., Zhang, S. and Tsukita, S., A novel amphioxus cadherin that localizes to epithelial adherens junctions has an unusual domain organization with implications for chordate phylogeny, *Evol. Dev.*, 4:426-434, 2002.

[16] Oda. H., Akiyama-Oda, Y., and Zhang, S., Two classic cadherin-related molecules with no cadherin extracellular repeats in the cephalochordate amphioxus: distinct adhesive specificities and possible involvement in the development of multicell- layered structures, *J. Cell Sci.*, 117:2757-2767, 2003.

[17] Page, R.D., TreeView: an application to display phylogenetic trees on personal computers, *Comput. Appl. Biosci.*, 12:357-358, 1996.

[18] Patthy, L., Modular assemlby of genes and the evolution of new functions, *Genetica*, 118:217-231, 2003.

[19] Putnam, N.H., Srivastava, M., Hellsten, U., Dirks, B., Chapman, J., Salamov, A., Terry, A., Shapiro, H., Lindquist, E., Kapitonov, V.V., Jurka, J., Genikhovich, G., Grigoriev, I.V., Lucas, S.M., Steele, R.E., Finnerty, J.R., Technau, U., Martindale, M.Q., and Rokhsar, D.S., Sea anemone genome reveals ancestral eumetazoan gene repertoire and genomic organization, *Science*, 317:86-94, 2007.

[20] Technau, U., Rudd, S., Maxwell, P., Gordon, P.M., Saina, M., Grasso, L.C., Hayward, D.C., Sensen, C.W., Saint, R., Holstein, T.W., Ball, E.E., and Miller, D.J., Maintenance of ancestral complexity and non-metazoan genes in two basal cnidarians, *Trends Genet.*, 21:633-639., 2005.

[21] Wang, M., and Gaetano-Anolles, G., Global Phylogeny Determined by the Combination of Protein Domains in Proteomes, *Mol. Biol. Evol.*, 23:2444-2454, 2006.

[22] `http://bioruby.net/`

[23] `http://dictybase.org/`

[24] `http://www.ensembl.org/`

[25] `http://flybase.bio.indiana.edu/`

[26] `http://hmmer.janelia.org/`

[27] `http://www.jgi.doe.gov/`

[28] `http://ncbi.nlm.nih.gov/`

[29] `http://ruby-lang.org/`

[30] `http://www.wormbase.org/`

A conservative parametric approach to motif significance analysis

Uri Keich, Patrick Ng

Department of Computer Science, Cornell University, Ithaca, NY, USA

We suggest a novel, parametric, approach to estimating the significance of the output of motif finders. Specifically, we rely on the good fit we observe between the 3-parameters Gamma family and the null distribution of motif scores. This fit was observed across multiple motif finders, background models and scoring functions. Under this parametric assumption we compute and show the utility of a conservative confidence interval for the p-value of the observed score. Since our method relies on the 3-parameters Gamma fit it should be applicable to a variety of finders.

1. Introduction

The identification of transcription factor binding sites, and of *cis*-regulatory elements in general, is an important step in understanding the regulation of gene expression. To address this need, many motif-finding tools have been described that can find short sequence motifs given only an input set of sequences. The motifs returned by these tools are evaluated and ranked according to some measure of statistical over-representation, the most popular of which is based on the information content or entropy [19] (see [3] for a recent comparative review).

Unfortunately, the area of motif significance analysis has lagged considerably behind the extensive development of tools for motif finding. Consider for example the popular profile or PWM (position weight matrix) based finders such as MEME [2], CONSENSUS [5] and the various approaches to Gibbs sampling (e.g. [8], [12], [6]). Many of these tools do not offer any significance evaluation at all * while others, notably MEME and CONSENSUS, rely on the E-value. Introduced originally in this context as the "expected frequency" [5] it is the expected number of random alignments of the same dimension that would exhibit an entropy score, or information content [19], that is at least as high as the score of the given alignment.

While a step in the right direction the E-value has significant shortcomings. First the E-values are currently only computed for the entropy score. The latter tacitly assumes the somewhat unrealistic, albeit popular, iid (independent and identically distributed) model for the background. The problem is that if the background sequences are generated using a more realistic model one quickly encounters

---

*Note that this problem differs from that of scanning sites with a known PWM, e.g., [20].

examples where ranking by the entropy score, or equivalently by the E-value, consistently yields sub-optimal motifs. Second, even when computed correctly [10] it can at times be conservative to the point where it is of no value [13]. This paper offers an alternative, parametric, approach to evaluate the significance analysis. Unlike the E-value, which only works for the entropy score assuming an iid model, our method works reasonably well for all combination of scores and background models we looked at.

Following the recommendation of [13] our new significance estimation takes into account the finder's specific performance. It hinges on an observation we previously made that the 3-parameter Gamma[†], or 3-Gamma for short, appears to fit very well F_f, the empirical distribution of the entropy score (under an iid background model) of several finders f. Here we first verified that this good fit extends to *every* combination of motif finder (MEME, CONSENSUS, many versions of Gibbs including Markov aware ones) and background model (iid or genomic samples) that we looked at.

Once we identify the parametric family of the null distribution of the score, F_f, we can use a parametric approach to evaluate the significance of the score. More precisely, by running our finder f on, say $n = 20$, random datasets (of the same dimension as the original input but from the assumed null distribution) we get a sample of size n from F_f. We can then fit a 3-parameter Gamma to this sample of scores and use the estimated parameters to obtain an estimation of the p-value of the observed score s. For the kind of parameters we are looking at (the shape parameter $a > 1$) this point estimator can be shown to be consistent [18] (i.e. it converges to the estimated p-value as $n \to \infty$). However this consistency is an asymptotic result and for small a sample size such as $n = 20$ this statistic can grossly over-estimate the significance of the observed score. One might be tempted to simply increase n to get a more reliable estimate, however that would increase the runtime of the finder by a factor of n. Thus in many cases we cannot realistically assume that we have access to a much larger sample size.

Here we complement the naive estimator by providing a conservative confidence interval for the estimated p-value. Conceptually we do so by first finding Θ, a 3-dimensional confidence set of the three estimated parameters of the distribution. We then maximize $1 - F_\theta(s)$ over all $\theta \in \Theta$ where $F_\theta(s)$ is the distribution function of a 3-Gamma with parameters θ and s is the observed score. While the idea itself is rather straightforward its implementation faces several difficulties which we had to overcome. In particular, how does one finds such 3-dimensional confidence sets and how can one guarantee a reasonably good maximization. These issues are discussed in more details below. The bottom line is that our significance evaluation is quite

---

[†]The distribution function of a 3-parameters Gamma with $\theta = (a, b, \mu)$ is a given by $F_\theta(s) = F_{\Gamma(a,b)}(s - \mu)$ where $F_{\Gamma(a,b)}$ is the Gamma distribution with it usual shape and scale parameters. We previously referred to it as a shifted-Gamma but a more standard definition is a 3-parameters Gamma where μ is the location parameter [7].

robust and should be applicable to many other combinations of scoring functions, motif finders and background models. To date no such general method is available except for a naive Monte Carlo estimation of the significance by directly comparing the observed score s to a random sample generated the same way we generate it. In the Section 7 below we compare our method with this as well as other significance evaluation methods.

2. The parametric family of the distribution of a motif finder's score

In [13] we showed that, especially in twilight zone motif searches, taking into account the performance of the specific motif finder that is considered yields a more reliable significance analysis. Thus, our new significance analysis attempts to estimate the distribution of the optimal score reported by the specific finder f which we denote F_f (or simply F). While in principle F can be estimated through extensive Monte Carlo simulations it is not in general a feasible solution. In particular F is not really a single distribution but rather an infinite number of distributions F^α, one per each set of values α of several relevant parameters such as: the size of the input sample, the finder's search parameters and the background model (whose parameters might be estimated from the sample). In practice this could mean we need to estimate from scratch a different F^α every time we run our finder which would present a generally unacceptable cost. However, if you know all the different distributions F^α belong to some parametric family then you only need to estimate the parameters of F^α rather than the entire distribution.

We previously mentioned that under the iid background model the null distribution of the optimal entropy score reported by several motif finders follows a 3-Gamma curve [13]. Interestingly, additional experiments we conducted using other scores[‡] as well as using the fairly realistic genomic background[§] indicate this result is fairly general: for all combinations of finders/scores and samples that we tested we found that the 3-Gamma family provides a fairly good fit to the empirical data. A couple of those fits are displayed in Figure 1.

At this point we do not know why the 3-Gamma family offers such good fits to these finder-specific distributions. Regardless, in what follows we assume these distributions do indeed belong to the 3-Gamma family. This is somewhat analogous to BLAST's assumption of a Gumbel distribution when analyzing gapped alignments. For while there exists an asymptotic theory for ungapped alignments no such theory can fully explain the observed Gumbel distribution in the gapped case [1]. Nevertheless, people rely on that assumption when assessing the significance of a gapped local alignment.

[‡]Specifically, ILR [13] and a generalization of the entropy score that adjusts for a Markov background.
[§]Where sequences were sampled out of a filtered human chromosome 22.

Fig. 1: Probability plots of 3-Gamma fits to the empirical score distribution

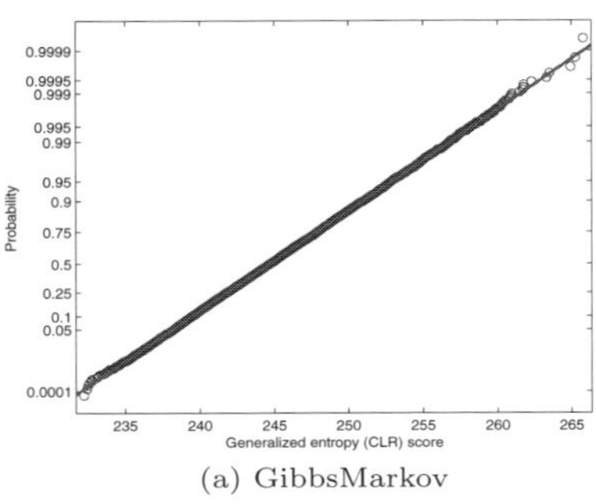

(a) GibbsMarkov

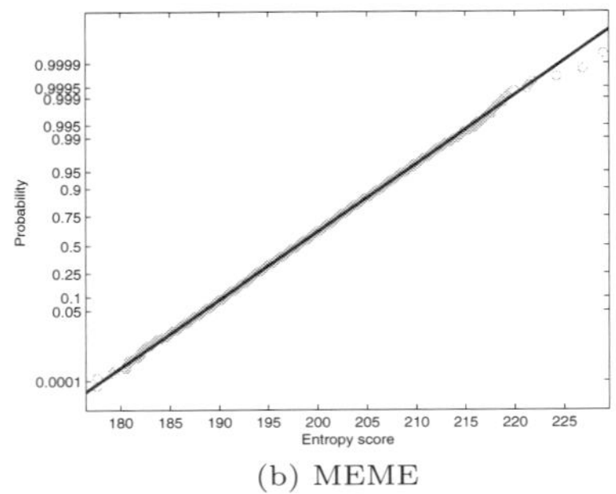

(b) MEME

The data was generated by running the mentioned program on 10^4 randomly generated datasets. In the case of MEME these were 20 sequences of each made of a block of 750 nucleotides sampled from a random location of a filtered human chromosome 22 whereas for GibbsMarkov we used 30 such sequences of length 1000 each

3. 3-Gamma based point estimator of the p-value

We previously showed that the parameters of the null distribution of CONSENSUS' entropy score can be reasonably grouped based only on two parameters of the finder [13]. In general however it is not clear how the parameters of the fitted distribution vary with the parameters of the problem (mentioned above as α). We therefore explore a different approach here relying on our ability to generate a *small* random sample $X = (X_1, \ldots, X_n)$ from F. Technically we generate this sample by first using the assumed background model to generate n independent random datasets of the same dimensions as the input dataset. We then run our finder on each of these n random datasets using the same settings as in the original application. Since this increases the runtime by a factor of n there is clear incentive to keep n as small as we can get away with.

Using this sample we can, for example, find $\hat{p}(s)$, the MLE (maximum likelihood estimator) of the p-value of the observed score s as follows. First we find $\hat{\theta} = \hat{\theta}(X)$, the MLE of the parameters of the 3-Gamma[¶], then we plug $\hat{\theta}$ into the definition of the p-value:

$$\hat{p}(s) = \hat{p}(s, X) = 1 - F_{\hat{\theta}}(s). \tag{1}$$

One problem is that with a small sample such as $n = 20$ one can only expect so much from fitting 3-parameters and indeed $\hat{p}(s)$ can badly over-estimate the significance of s (see Figure 2a). A standard way to account better for the variability of the sample is to introduce confidence intervals (e.g. [17]). For example, a 90%

[¶]In [13] we suggested fitting a 3-Gamma relying on our ability to fit the standard Gamma distribution. Here we adopt a different strategy relying on our ability to express the location parameter μ in closed form in terms of the shape and scale parameters at a critical point (calculation not shown).

confidence interval for the p-value is a random interval that contains the real p-value $p(s)$ with probability ≥ 0.9. We next construct such a confidence interval.

4. Confidence set for θ_0, the 3-Gamma parameters

Conceptually we construct our confidence interval for $p(s)$, the p-value of the observed score s in two steps. First, as described next, we use a generalization of the profile likelihood method (e.g., [14]) to construct a 3-dimensional confidence set for the 3-Gamma parameters θ_0. We then scan this confidence set to maximize the p-value of s at the prescribed confidence level.

Let $X = (X_1, \ldots, X_n)$ be our random sample and let $\hat{\theta} = \hat{\theta}(X)$ be the 3-Gamma MLE, so $\hat{\theta}$ is a 3-dimensional random vector. Let $L(X; \theta)$ be the 3-Gamma log-likelihood of the sample given the parameter $\theta \in \Omega$, where $\Omega \subset \mathbb{R}^3$ is the set of feasible parameters which would generally be a subset of $(a > 1^{\|}, b > 0, \mu \in \mathbb{R})$.

According to a general asymptotic result if the sample X is drawn according to the 3-Gamma distribution F_{θ_0} then

$$\Delta_L(X; \theta_0) = 2 \left[L(X; \hat{\theta}) - L(X; \theta_0) \right]$$

converges in distribution to a $\chi^2(3)$ distribution as n, the size of the sample X, goes to infinity (e.g., [17]). *Assume* for now that this asymptotic result holds for our finite sample and define

$$\Theta_\gamma(X) = \left\{ \theta \in \Omega \, : \, \Delta_L(X; \theta) \leq F^{-1}_{\chi^2(3)}(\gamma) \right\}, \tag{2}$$

where $F^{-1}_{\chi^2(3)}(\gamma)$ is the γ-quantile of the $\chi^2(3)$ distribution. Thus, $\Theta_\gamma(X)$ is random subset of $\Omega \subset \mathbb{R}^3$.

Claim 4.1. *Assuming $\Delta_L(X; \theta_0)$ is distributed under θ_0 as $\chi^2(3)$, $\Theta_\gamma(X)$ is a γ-confidence set for θ_0.*

Proof. We need to show that for any $\theta_0 \in \Omega$, $P_{\theta_0}(\theta_0 \in \Theta_\gamma(X)) \geq \gamma$. This follows immediately from the definitions:

$$P_{\theta_0}(\theta_0 \in \Theta_\gamma(X)) = P_{\theta_0}\left[\Delta_L(X; \theta_0) \leq F^{-1}_{\chi^2(3)}(\gamma) \right] = F_{\chi^2(3)}\left[F^{-1}_{\chi^2(3)}(\gamma) \right] = \gamma. \qquad \square$$

When constructing confidence intervals it is a standard practice to assume as we did in the last claim, that an asymptotic distribution holds for a finite sample. Nevertheless, we would like to test what kind of errors does this assumption introduce in our case. This is particularly important since for practical reasons we restrict Ω, the set of feasible parameters, so that the shape parameter (coordinate) is restricted to a certain interval. We therefore conducted the following experiment.

$^{\|}$ We need $a > 1$ to guarantee the success of the estimation process as well as for the asymptotic result below [18]. Fortunately this is not a real restriction in our case as the distributions we are interested in have $a \gg 1$.

We first generated 44 different large sets of empirical scores. In this case all the scores came from one finder, GibbsMarkov, which is our version of Gibbs Sampler that uses a variant of the entropy score that accounts for a higher order Markov background model. GibbsMarkov will be described in detail in a following paper but for now it suffices to say it is similar to BioProspector [9]. Each such set of empirical scores contained 10^4 applications of GibbsMarkov on that many randomly generated datasets (with fixed dimensions per set). The sequences were randomly sampled from a filtered human chromosome and the dimensions ranged from 10 sequences of length 750 to 30 sequences of length 1000 each. Similarly, the width of the motif searched by GibbsMarkov varied from 8 to 28.

We then estimated the 3-Gamma parameters θ_0 for each of these 44 sets of empirical scores and defined $\Omega = \{(a, b, \mu) : a \in [10, 100], b > 0, \mu \in \mathbb{R}\}$. This range for the shape parameters was determined by taking a slightly larger interval than necessary to contain all 44 estimated shapes. Finally, we forget about the original sets and simply generate a large number (10^4) of random 3-Gamma samples of size $n = 20$ for each of these 44 estimated parameters θ_0. Since we know θ_0 in this case we can readily determine the proportion of times $\theta_0 \in \Theta_\gamma(X)$ and compare it to the theoretical rate of γ. The results of this test are summarized in Table 1.

Table 1: Actual confidence coefficients of parameter sets

γ in (2) set to:	0.85	0.90	0.95	0.99
actual confidence (%):	89.4-90.7	92.9-94.0	96.6-97.3	99.3-99.6

Range reported is of the percentage of time $\theta_0 \in \Theta_\gamma(X)$ for the specified γ observed across the 44 tests. Note how stable the observed ranges are, allowing us to correct for the slight conservative bias of the original $\chi^2(3)$ derived thresholds.

The table demonstrate our confidence sets are consistently slightly conservative. Consulting this table we can however adjust for the conservative nature of these confidence sets: for example, a nominal 85% confidence set is in fact a 90% one. Regardless of whether or not we adopt this adjustment we next show how we use our confidence set to generate a confidence interval for our real object of interest: the p-value of s, $p(s) = 1 - F_{\theta_0}(s),$.

5. 3-Gamma based confidence interval for the *p*-value

Let

$$\hat{p}_c = \hat{p}_c(s, X) = \max\left\{1 - F_\theta(s) : \theta \in \Theta_\gamma(X)\right\}, \tag{3}$$

where F_θ is the 3-Gamma distribution with parameter θ.

Claim 5.1. *The random interval $[0, \hat{p}_c(s, X)]$ is a confidence interval for the p-value of s, $p(s)$, with confidence coefficient $\geq \gamma$.*

Proof. Since $\Theta_\gamma(X)$ is a γ-confidence set, $\theta_0 \in \Theta_\gamma(X)$ with probability $\geq \gamma$. In this case we clearly have

$$F_{\theta_0}(s) \geq \min\{F_\theta(s) : \theta \in \Theta_\gamma(X)\} \tag{4}$$

and therefore

$$p(s) = 1 - F_{\theta_0}(s) \leq 1 - \min\{F_\theta(s) : \theta \in \Theta_\gamma(X)\} = \hat{p}_c(s, X).$$

Thus, $p(s) \in [0, \hat{p}_c(s, X)]$ with probability $\geq \gamma$. $\qquad\square$

Comment. Note that this estimate is conservative in nature as (4) might often hold even when $\theta_0 \notin \Theta_\gamma(X)$.

While conceptually our method for generating the confidence interval for $p(s)$ works as described above, we found that technically it is better to combine the two steps into one. More precisely, we define a target function for maximization:

$$\varphi(\theta; X, s) = \begin{cases} 1 - F_\theta(s) & \Delta_L(X; \theta) \leq d \\ -\Delta_L(X; \theta) & \Delta_L(X; \theta) > d \end{cases},$$

where $d = d_\gamma = F^{-1}_{\chi^2(3)}(\gamma)$. The following claim guarantees it suffices to maximize $\varphi(\theta; X, s)$:

Claim 5.2. $\max_{\theta \in \Omega} \varphi(\theta; X, s) = \hat{p}_c(s, X)$

Proof. Immediate from the fact that $\varphi(\theta; X, s) < 0$ for $\theta \notin \Theta_\gamma(X)$. $\qquad\square$

Mathematically, $\max_{\theta \in \Omega} \varphi(\theta; X, s) = \hat{p}_c(s, X)$ is a well defined statistic which, assuming the validity of the χ^2 approximation, defines a γ-confidence interval for the p-value, $p(s)$. However, in practice maximizing φ over Ω turned out to be somewhat tricky as the landscape of φ defined on $\Omega \subset \mathbb{R}^3$ is apparently quite complicated. This means that the actually computed version of $\hat{p}_c(s, X)^{**}$ might yield a confidence interval that would be smaller than it should be, i.e., its confidence coefficient would be $< \gamma$. In this case we would fail to achieve our goal here to get a confidence interval for $p(s)$ (we already have a reasonable point estimate in $\hat{p}(s)$ defined in (1) above).

In practice we used the Nelder-Meade [11] simplex based optimization procedure (implemented in the `constrOptim` function in R [15]) to maximize $\varphi(\theta; X, s)$ over $\theta \in \Omega$. Using Monte Carlo simulations for which we know the correct p-value we learned that simply relying on multiple random restarts is not satisfactory with this as well as with gradient based optimizations.We therefore used a pre-defined lattice of "reasonably good" starting points next to the boundary of Ω. Figure 2 shows a typical histogram of the computed conservative $\hat{p}_c(s, X)$ compared with the point estimator $\hat{p}$.

**We abuse the notations by not distinguishing between the mathematically defined statistic and its computed version.

Fig. 2: Comparing the estimators $\hat{p}$ and $\hat{p}_c(s, X)$ of $p = 10^{-3}$

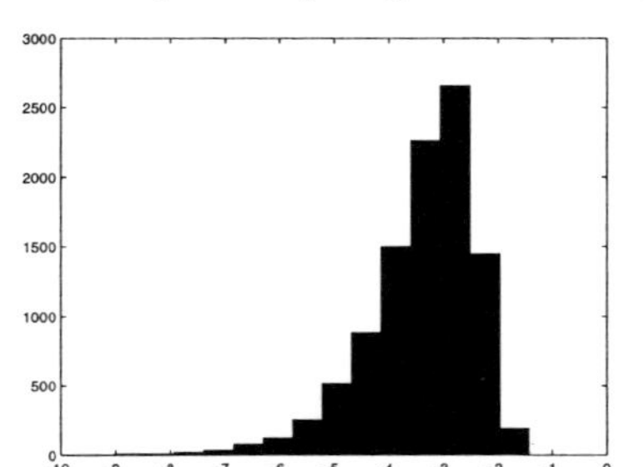

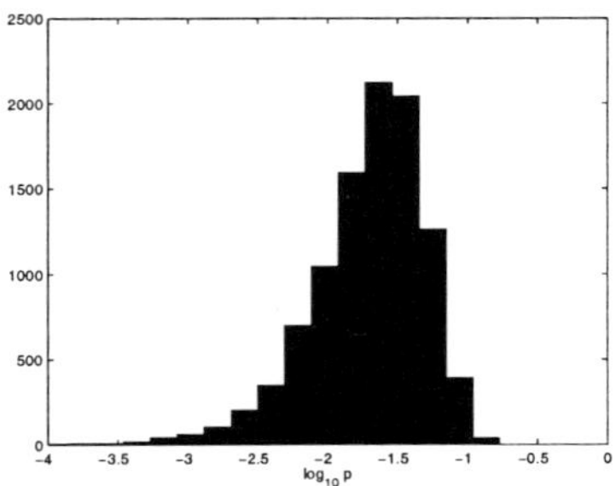

(a) $\hat{p}$ can overestimate the significance

(b) $\hat{p}_c(s, X)$ is mostly conservative

Histograms of 10^4 independent evaluations of the point estimator $\hat{p}$ of $s = F_{\theta_0}^{-1}(10^{-3})$ and of the conservative $\hat{p}_c(s, X)$. The necessary 10^4 samples of size $n = 20$ were drawn with repetitions from a large set of empirical scores of the finder. In this case the set was made of 10^4 runs of GibbsMarkov on that many randomly generated datasets, each of which consisting of 30 sampled genomic sequences of length 1000. Since θ_0 is the 3-Gamma MLE of the set of empirical scores from which the samples of size n were taken, by definition, the real p-value of $s = F_{\theta_0}^{-1}(10^{-3})$ is essentially 10^{-3}.

6. The fidelity and utility of the confidence interval for the p-value

It was reassuring to see above that our confidence sets for θ_0 attain their prescribed confidence level. However, especially in light of the difficulty in maximizing φ over Ω, the more important question is whether or not our computed confidence interval $[0, \hat{p}_c(s, X)]$ contains $p(s)$ with probability $\geq \gamma$. To test that we conducted the following experiment based on our 44 sets of empirical scores described above.

We chose a set of p-values ranging from 10^{-9} to 0.1 and computed the values $s = s(\theta_0, p_0)$ for which $F_{\theta_0}(s) = p_0$[tt]. Again, θ_0 is the 3-Gamma MLE obtained by fitting a 3-Gamma to each of these 44 empirical scores sets. For each of these s_0 (one per p-value and empirical score set) we computed $\hat{p}_c(s_0, X)$ for 10^4 samples X of $n = 20$ scores drawn independently, with repetitions, from the appropriate empirical set of scores. Note that when computing $\hat{p}_c(s_0, X)$ one assumes a specific confidence coefficient γ. We could then find the percentage of time $p_0 \in [0, \hat{p}_c]$ and test whether or not it is bigger than the prescribed γ. Table 2 gives the positive summary for all the cases we looked at.

By construction $\hat{p}_c(s, X)$ is a conservative estimate of $p(s)$ so we should not be surprised that, as observed above, it tends to underestimate the significance of s. We should however expect that it would not loose all the information. To demonstrate the utility of $\hat{p}_c(s, X)$ we conducted two tests as extensions of the

---

[tt]Except for $p_0 \geq 0.005$ for which we could fairly reliably estimate s_0 directly from the empirical distribution of scores.

Table 2: Fidelity of confidence coefficient of estimated p-values

p_0	0.1	0.01	10^{-3}	10^{-4}	10^{-6}	10^{-9}
median %	1.70	1.69	1.23	1.36	1.16	0.85
minimum %	1.00	0.65	0.30	0.40	0.28	0.23
maximum %	2.55	5.22	3.50	3.77	4.38	4.67

γ in (2) was set to 0.85 which per Table 1 should really be a 90% confidence coefficient. The first row gives the prescribed p-value p_0. Rows 2-4 give the median, minimum and maximum of the percentage of samples for which $p_0 \notin [0, \hat{p}_c]$ among all 44 sets (a percentage for each set was computed as described in the text). Note that the even the worst case scenarios still attain the prescribed γ. The results for $\gamma = 0.9$ were qualitatively the same.

previously described test. In the first of these tests we asked for the percentage of samples X above for which $\hat{p}_c(s, X) \leq 0.05$. The latter represent an analogue of the canonical 5% significance threshold. More interesting is the actual value of $\hat{p}_c(s, X)$ so we looked at its median value across all 10^4 samples X. Table 3 summarizes the fluctuations of these statistics across all 44 empirical sets as a function of the actual p-value. Note how, especially for small p-values, $\hat{p}_c$ conveys significantly more information than simply "I passed the 5% significance threshold". For example, for $p_0 = 10^{-6}$ the median value (over all 44 sets) of the median (over all samples X) of $\hat{p}_c$ is roughly 6e-4.

Table 3: The utility of the confidence interval for the p-value

p_0	0.1	0.01	10^{-3}	10^{-4}	10^{-6}	10^{-9}
median %	0.10	32.33	86.67	99.40	100.00	100.00
minimum %	0.03	23.05	81.53	98.55	99.97	100.00
maximum %	0.30	47.60	92.90	99.85	100.00	100.00
median $\hat{p}_c$	0.25	0.068	0.022	0.0067	0.00059	1.2e-05
minimum $\hat{p}_c$	0.24	0.052	0.015	0.0035	0.00017	1.4e-06
maximum $\hat{p}_c$	0.26	0.079	0.027	0.0094	0.0012	3.9e-05

γ in (2) was set to 0.85 (90% in practice). The first row gives the prescribed p-value p_0. Rows 2-4 give the median, minimum and maximum of the percentage of samples for which $\hat{p}_c \leq 0.05$ among all 44 sets. Rows 3-6 yield the median, minimum and maximum among all 44 medians of $\hat{p}_c$. The results for $\gamma = 0.9$ were qualitatively the same.

One should keep in mind that with larger n the accuracy of $\hat{p}_c(s, X)$ can improve significantly. For example we compared using a sample of size $n = 20$ to $n = 40$ for $p_0 = 10^{-6}$. We found that while $n = 20$ yields a median (of medians) of roughly 5.9e-4, using a sample of size $n = 40$ cut the median to roughly 1.6e-4.

For a final practical test we went back to the Gibbs Sampler results on the COMBO experiment from [13] for which the E-value assessment failed miserably: the median of the *positive* examples was $\approx 10^{12}$. Using a set of 1600 runs on null iid datasets of the same dimension as in the original experiment we generated samples of size $n = 20$ and computed $\hat{p}_c(s, X)$ for each of the 400 scores s (each s is the entropy score of the Gibbs Sampler applied to a different, implanted dataset; see [13] for details).

We *predicted* a run as positive or successful if $\hat{p}_c(s, X) \leq 0.05$ and negative, or failure otherwise. We *labeled* a run as positive if the overlap between the reported and implanted alignment was $\geq 30\%$ and negative otherwise. Thus, we could count the number of TPs and FPs. To smooth out the results we repeated this process 100 times and averaged the number of TPs and FPs. Using $\gamma = 0.85$ our classifier defined above averaged 140.1 TPs and 6.3 FPs and an average of 55.6 of the scores s had a much more significant $\hat{p}_c(s, X) \leq 0.01$. Moving to a larger sample size of $n = 40$ we averaged 168.4 TPs and 8.5 FPs and an average of 80.6 had $\hat{p}_c(s, X) \leq 0.01$.

7. Discussion

We presented a novel approach for evaluating the significance of a motif finder results. It is important to keep in mind that as long as the fit of the finder's empirical scores distribution to a 3-Gamma is a reasonable one our method should be applicable to that finder. Since we have yet to see a case where that fit is not good we believe our method should apply to a wide variety of combinations of motif finders and scores and thus offer a unified parametric approach to estimating a finder's specific performance[‡‡].

We should point out that while our method suffers from a time penalty factor of n (the sample size), computing $\hat{p}_c(s, X)$ can be readily executed in parallel so that if sufficient additional CPU cores are available the effective time penalty reduces to only a factor of 2.

What are alternative significance evaluations? The authors of GLAM [4] assume that a scoring function they derive has a Gumbel distribution. They then try to evaluate its parameters analogously to BLAST. In particular, similarly to the E-value calculation they do not require the costly "on-the-fly" generation of a sample of null scores. While their method works reasonably well for a small number of sequences (≈ 5) it seems that the Gumbel assumption fails for a larger, more typical, number of sequences.

A more general alternative to the 3-Gamma distribution is the generalized extreme value (GEV) distribution [16]. While both of these distribution families offer fairly close fits and are difficult to distinguish at times, we found that typically the 3-Gamma offers a more reliable prediction of the right tail which is the one we are interested in. Additionally the 3-Gamma family was easier to handle in terms of

---

[‡‡]Technically, adjusting our method to a new finder would typically require some crude charting of the space of plausible values for the parameters: the more restrictive the range of feasible parameters Ω is, the more accurate will $\hat{p}_c(s, X)$ be.

fitting and predicting confidence sets. Finally, while the GEV might sound attractive as it is known to be the only possible asymptotic limit of a maximum of an iid sequence one should keep in mind that the motif finding problem is very different from the alignment problem where such extreme value theory applies.

Although we could not find any trace of this in the program itself, the Bio-Prospector paper [9] suggests an approach which is similar in spirit to the computation of our point estimator $\hat{p}$ (1). One major difference is they suggest that the normal approximation should be used. We found no evidence supporting the use of a normal approximation and no such convincing evidence is presented in that paper. In all our studies the 3-Gamma family offered significantly superior fits at a cost of only one more parameter to estimate.

Alternatively we can resort to non-parametric tests. For example, we can use the generated sample to construct confidence intervals for the p-value the same way we estimate p of a binomial $B(n,p)$ distribution. The problem is these tend to be quite conservative so for $n = 20$ the best confidence interval for the p-value would be $[0, 0.11]$ while for $n = 40$ it would be $[0, 0.06]$.

A different kind of non-parametric test is to test, for example, if s is bigger than all the entries in a random sample of size $n = 20$ (this can be generalized using the Mann-Whitney statistic). As described this is a reliable test at the (roughly) 5% significance level whose main down side is that it offers very little information about the quality of significant results. In particular, it will not provide any more information about a score whose real p-value is 10^{-6} than it would about any other score s that passes the 5% test: all you learn is that you are 95% confident that the observed dataset is not a random one. Our method on the other hand, though conservative, does respond to differences between scores which among other things would make it more appropriate to compare motifs of different widths where the scores cannot be compared directly against one another. Moreover, this non-parametric method has a high "false positive" rates for scores s whose p-value is close to 0.05. For example, if $p(s) = 0.1$ then 12% of the time s will be declared significant at the 5% level and if $p(s) = 0.06$ this will happen 29% of the time.

There are many directions and questions that our paper opens up including: applying our significance analysis method to other finders and making them an integral option of GibbsMarkov as well as other finders, developing a better theoretical understanding of why the 3-Gamma offers such good fits to the optimal score distributions, and testing how good a fit remains once we start adding additional information such as ChIP-chip or phylogeny data to our motif finders. We plan on exploring these issues in future research.

References

[1] SF Altschul and W Gish. Local alignment statistics. *Methods Enzymol*, 266:460–80, 1996.

[2] T.L. Bailey and C. Elkan. Fitting a mixture model by expectation maximization to discover motifs in biopolymers. In *Proceedings of the Second International Conference*

on *Intelligent Systems for Molecular Biology*, pages 28–36, Menlo Park, California, 1994.

[3] Martin Tompa et al. Assessing computational tools for the discovery of transcription factor binding sites. *Nat Biotechnol*, 23(1):137–44, Jan 2005.

[4] Martin C Frith, Ulla Hansen, John L Spouge, and Zhiping Weng. Finding functional sequence elements by multiple local alignment. *Nucleic Acids Res*, 32(1):189–200, 2004.

[5] GZ Hertz and GD Stormo. Identifying DNA and protein patterns with statistically significant alignments of multiple sequences. *Bioinformatics*, 15(7-8):563–77, 1999.

[6] JD Hughes, PW Estep, S Tavazoie, and GM Church. Computational identification of cis-regulatory elements associated with groups of functionally related genes in Saccharomyces cerevisiae. *J Mol Biol*, 296(5):1205–14, Mar 2000.

[7] Kotz S. Johnson N.L. and Balakrishnan N. *Continuous Univariate Distributions, 2nd edition*. Wiley Series in Probability and Statistics, 1994.

[8] CE Lawrence, SF Altschul, MS Boguski, JS Liu, AF Neuwald, and JC Wootton. Detecting subtle sequence signals: a Gibbs sampling strategy for multiple alignment. *Science*, 262(5131):208–14, Oct 1993.

[9] X Liu, DL Brutlag, and JS Liu. BioProspector: discovering conserved DNA motifs in upstream regulatory regions of co-expressed genes. *Pac Symp Biocomput*, pages 127–38, 2001.

[10] Niranjan Nagarajan, Neil Jones, and Uri Keich. Computing the P-value of the information content from an alignment of multiple sequences. *Bioinformatics*, 21 Suppl 1(ISMB 2005):i311–i318, Jun 2005.

[11] J. A. Nelder and R. Mead. A simplex algorithm for function minimization. *Computer Journal*, 7:308?313, 1965.

[12] AF Neuwald, JS Liu, and CE Lawrence. Gibbs motif sampling: detection of bacterial outer membrane protein repeats. *Protein Sci*, 4(8):1618–32, Aug 1995.

[13] Patrick Ng, Niranjan Nagarajan, Neil Jones, and Uri Keich. Apples to apples: improving the performance of motif finders and their significance analysis in the Twilight Zone. *Bioinformatics*, 22(14):e393–401, Jul 2006.

[14] Yudi Pawitan. A reminder of the fallibility of the wald statistic: Likelihood explanation. *he American Statistician*, 54(1):54–56, 2000.

[15] R Development Core Team. *R: A Language and Environment for Statistical Computing*. R Foundation for Statistical Computing, Vienna, Austria, 2006. ISBN 3-900051-07-0.

[16] Sidney I. Resnick. *Extreme values, regular variation, and point processes*. Springer-Verlag, New York, 1987.

[17] J.A. Rice. *Mathematical Statistics and Data Analysis*. Duxbury Press, second edition, 1995.

[18] Richard L. Smith. Maximum likelihood estimation in a class of nonregular cases. *Biometrika*, 72(1):67–90, 1985.

[19] GD Stormo. DNA binding sites: representation and discovery. *Bioinformatics*, 16(1):16–23, Jan 2000.

[20] Jing Zhang, Bo Jiang, Ming Li, John Tromp, Xuegong Zhang, and Michael Q. Zhang. Computing exact P-values for DNA motifs. *Bioinformatics*, 23(5):531–537, 2007.

RECOGNITION OF POLYADENYLATION SITES FROM ARABIDOPSIS GENOMIC SEQUENCES

CHUAN HOCK KOH   LIMSOON WONG
kohchuan@comp.nus.edu.sg   wongls@comp.nus.ed.sg

School of Computing, National University of Singapore
COM1, Law Link, Singapore 117590

A polyadenine tail is found at the 3' end of nearly every fully processed eukaryotic mRNA and has been suggested to influence virtually all aspects of mRNA metabolism. The ability to predict polyadenylation site will allow us to define gene boundaries, predict number of genes present in a particular gene locus and perhaps better understand mRNA metabolism. To this end, we built an arabidopsis polyadenylation prediction model. The prediction model uses a machine learning method which consists of four sequential steps: feature generation, feature selection, feature integration and cascade classifier. We have tested our model on public datasets and achieved more than 97% sensitivity and specificity. We have also directly compared with another arabidopsis prediction model, PASS 1.0, and have achieved better results.

Keywords: arabidopsis, machine learning, polyadenylation site

1.   Introduction

Polyadenylation is a post-transcriptional process. The process basically cleaves and adds about 200 adenosine residues to the pre-mRNA 3' end. The site where the pre-mRNA is cleaved is known as the polyadenylation site. The selection of polyadenylation sites are determined by polyadenylation signals or cis-elements in the pre-mRNA. In humans, AAUAAA is a highly conserved polyadenylation signal. However, no highly conserved polyadenylation signal has been identified in arabidopsis. In this respect, the prediction of polyadenylation site is therefore more difficult.

The polyadenine tail has been shown to boost translation, protects the 3' end of mRNA from exonucleases and is needed for the mRNA nuclear-to-cytoplasmic export. This process has also been found to be tightly coupled with splicing and transcription termination. Thus, it is an essential processing event and an integral part of gene expression [5].

Therefore, the ability to predict the polyadenylation site potentially allows us to better understand the process and also to be able to better segment genes. To this end, we developed an arabidopsis polyadenylation prediction model.

Although proteins involved in polyadenylation processes appears to be conserved among human and arabidopsis, their polyadenylation signals differ widely in terms of their locations with respect to polyadenylation site and sequence content [5]. In humans, AAUAAA (or its one-base variant) is a highly conserved polyadenylation signal and is found in 87.1% of the observed sites [1]. In plants, there are no highly conserved polyadenylation signals. AAUAAA is the most frequently occurring polyadenylation signal, yet it is found in only about 10% of arabidopsis genes [5].

Nonetheless, studies have shown that arabidopsis polyadenylation signals are composed of three major groups: far upstream elements (FUE), near upstream elements (NUE) and cleavage elements (CE). The far upstream elements span a region of 60-130 nucleotides, resides at a location 18 to 22 nucleotides upstream of the cleavage site and has a high U content. The near upstream elements spans a region of 6-10 nucleotides, resides at a location 12 nucleotides upstream of the cleavage site and has a high A content. The cleavage elements span a region of about 27 nucleotides, resides from a position which is 12 nucleotides upstream to about 15 nucleotides downstream of the cleavage site and it consists of the well conserved YA(CA or TA) just before the cleavage site and U-rich elements flanking both sides of the cleavage site [5]. For a schematic representation of polyadenylation site in arabidopsis mRNA 3' ends, see Figure 1.

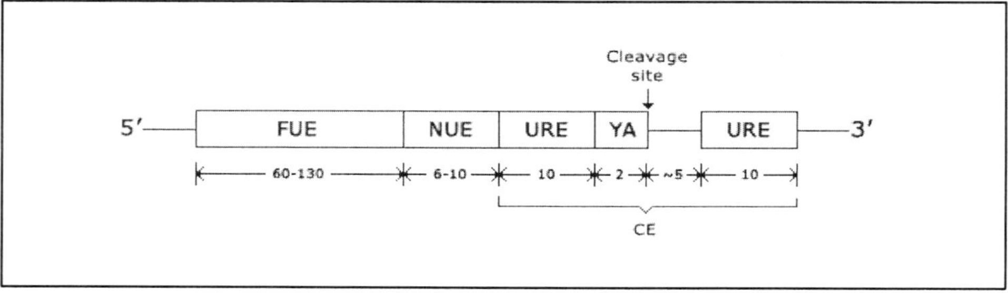

Figure 1. A schematic representation of polyadenylation site in arabidopsis mRNA 3' ends.

Currently, there exists another prediction program for arabidopsis named polyadenylation site sleuth or PASS [3]. PASS was developed based on Generalized Hidden Markov Model using polyadenylation signals identified by [5]. We compared against PASS 1.0 using the same datasets and achieved better results in most cases.

For our arabidopsis polyadenylation prediction, we built a model based on the machine learning methodology described in [4]. However, we added an additional step to that methodology - a cascade classifier. This additional step is implemented due to the fact that there are no highly conserved polyadenylation signals in Arabidopsis; and this step has been shown to increase both sensitivity and specificity.

2.   Tools, Datasets, and Methods

2.1 *Tools*

In our prediction model, certain steps were implemented using Waikato Environment for Knowledge Analysis (WEKA). WEKA is a free machine learning software package written in Java and developed at University of Waikato [8]. One particular learning scheme that we have employed from WEKA is SMO. SMO is the WEKA implementation of support vector machine [7] using John Platt's sequential minimal optimization algorithm [6].

2.2 *Datasets*

Datasets used for this project came from two main sources.
1) Datasets derived by Hao Han
- Dataset A (used to set parameters):
    - o 804 (+ve) sequences with EST-supported polyadenylation sites, derived based on ATPACDB and has confidence "high" or "very high". (http://harlequin.jax.org/atpacdb/confidence.php)
    - o 9742 (-ve) coding sequences that were extracted from ENSEMBL database, arabidopsis section
    - o Sequences in Dataset A are of length 400. For each of the 804 (+ve) sequences, the EST-supported polyadenylation is at location 201.

2) Datasets provided by Qingshun Li; please refer to [3] for more information on how the following datasets were derived
- Dataset B (used for SMO1 training):
    - o 2640 (+ve) sequences with EST-supported polyadenylation sites
    - o 900 (-ve) coding sequences
    - o 476 (-ve) 5'UTR sequences
    - o 954 (-ve) intronic sequences
- Dataset C (used for SMO2 training):
    - o 1500 (+ve) sequences with EST-supported polyadenylation sites
    - o 100 (-ve) coding sequences
    - o 100 (-ve) 5'UTR sequences
    - o 100 (-ve) intronic sequences
- Dataset D (used for SMOA and SMO2 testing):
    - o 2069 (+ve) sequences with EST-supported polyadenylation sites
    - o 501 (-ve) coding sequences
    - o 288 (-ve) 5'UTR sequences
    - o 527 (-ve) intronic sequences
- Dataset E (used for SMOA training):
    - o 4140 (+ve) sequences with EST-supported polyadenylation sites
    - o 1000 (-ve) coding sequences
    - o 576 (-ve) 5'UTR sequences
    - o 1054 (-ve) intronic sequences
- Each sequence in Dataset B, C, D and E is of length 400. Each (+ve) sequences has the EST-supported polyadenylation sites at location 301. Dataset E is formed by combining Dataset B and Dataset C. Each sequence in Dataset B, C and D underwent pair-wise global alignment against every other sequence. If any two sequences have more than 70% similarity, one of them is discarded. (i.e., if we

randomly pick a sequence from Dataset B, we will not find any other sequence in Dataset B, C or D with more than 70% similarity). This is to minimize biasness due to similarity of sequences.

2.3 *Methods*

2.3.1 *Architecture of the polyadenylation prediction system*

Our polyadenylation prediction system is a cascade of two layers of classifiers. In the first layer, a classifier SMO1 is used to score positions (-40/+40) at every nucleotide and therefore, there is a total of 81 SMO1 scores relative to a candidate site. In the second layer, a classifier SMO2 takes these 81 SMO1 scores to decide if the candidate site is a polyadenylation site.

We follow the "feature generation, feature selection, feature integration" methodology [4] in developing our prediction system, and in particular, the first-layer classifier SMO1.

The architecture of our polyadenylation prediction system and an overview of the steps involved are depicted in Figure 2.

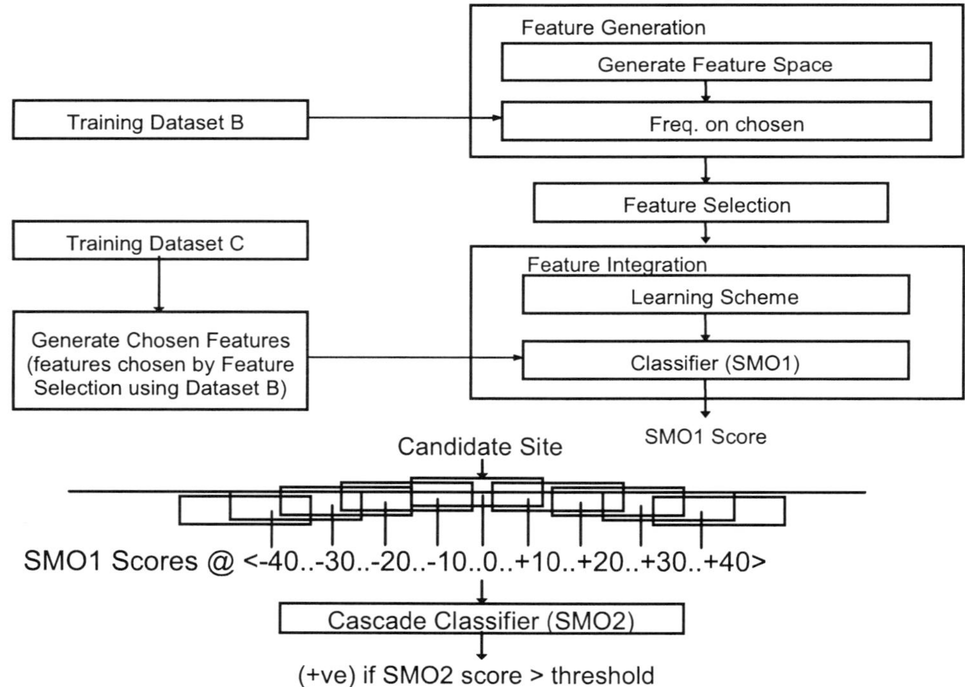

Figure 2. Architecture of our polyadenylation prediction system.

2.3.2 *Training Phase*

1) Feature Generation

The candidate features that we considered for SMO1 are 1-gram (A, C, G, U), 2-gram (AA, AC, AG, AU,..., GU, UU), 3-gram (AAA, AAC, AAG, AAU, ..., UGU, UUU), 4U/1N (NUUUU, UNUUU, UUNUU, UUUNU, UUUUN), 4A/1N (NAAAA, ANAAA, AANAA, AAANA, AAAAN) and G/U*7 (A stretch of G or U for 7 bp). We consider these features separately in 3 different windows relative to the candidate site, (-110/+5), (-35/+15) and (-50/+30); and we calculate the frequency of the features in these 3 different windows. A total of 261 candidate features were generated.

We chose these features and windows by referencing to biological literature [1, 2 and 5] and findings from analyzing Dataset A.

2) Feature Selection

The feature selection step is done by using WEKA supervised attribute filter with Attribute Evaluator set to "ChiSquaredAttributeEval" and Search Method set to "Ranker". Statistics of these candidate features are then computed based on Dataset B for SMO1. Features that have chi-square statistics greater than the threshold value of 0 are chosen to form the feature vector.

Out of the 261 candidate features generated, 228 have chi-square value exceeding 0 and were selected. Of them, the minimum observed chi-square value is 11.

3) Feature Integration

The first-layer classifier SMO1 is then trained using the SMO support vector machine learning scheme in WEKA and the selected features from Step 2. The training data used in this step is from Dataset B.

4) Cascade Classifier (SMO2)

The second layer classifier SMO2 is then trained using the SMO support vector machine learning scheme in WEKA. The feature vector for this step is the 81 scores output by SMO1 at positions (-40/+40) relative to a candidate site. The training data for this step is from Dataset C.

2.4 *Prediction Phase*

Although in Figure 1, "YA" is said to be well conserved at positions just before the polyadenylation sites, it is found in less than 41% of the sequences in Dataset A. Therefore, when given a sequence to predict the existence and location of polyadenylation sites, we consider every location to be a possible candidate site instead of only "YA" positions. Hence, the classifier SMO1 is first deployed to get a score at every nucleotide for a given sequence. Cascade classifier SMO2 then makes use of the SMO1 scores at positions (-40/+40) relative to a candidate site to carry out prediction for polyadenylation sites.

However, in order to make predictions on a particular nucleotide, SMO1 needs to know that particular nucleotide's (-110/+30) composition. Therefore, for SMO2 to make a prediction, it needs to know the composition of (-150/+70). Hence, given a sequence of DNA, our prediction model is not able to make prediction on the first 150 and last 70 nucleotides of the sequence. Likewise, PASS 1.0 is also unable to make predictions on the first 149 and last 10 nucleotides. PASS 1.0 assumes these locations to be unlikely to be a polyadenylation site and sets the scores for these locations to be 0. Our prediction model also assumes these locations to be unlikely to be a polyadenylation site and sets their scores to 0. Therefore, the users of our model and/or PASS 1.0 are advised to provide sequences that are of sufficient length. (i.e., having upstream and downstream of a sequence to go beyond 3'UTR).

3. Results

In order to show improvements given by using the cascade classifier step, we also trained another classifier SMOA with Dataset E using the same steps for SMO1. Using Dataset E for SMOA ensures that SMOA and SMO2 uses equal amount of training data.

We then used Dataset D to evaluate the performance of our polyadenylation prediction system (SMO2), SMOA and PASS 1.0. We considered the following measures:

$$\text{Sensitivity (SN)} = TP / (TP + FN)$$
$$\text{Specificity (SP)} = TN / (TN + FP)$$

where TP (True Positive) is the total number of EST-supported polyadenylation sites that are identified or correctly predicted in the (+ve) sequences. FN (False Negative) is the total number of EST-supported polyadenylation sites that are not identified or predicted correctly in the (+ve) sequences. TN (True Negative) is the total number of sites with score $\leq$ threshold in the (-ve) sequences. FP (False Positive) is the total number of sites with score $>$ threshold in the (-ve) sequences.

The sensitivity and specificity of SMOA, SMO2 and PASS 1.0 running on the test dataset can be observed in Figures 3, 4 and 5 respectively. The equal-error-rates values (i.e., the points where sensitivity = specificity) achieved by SMOA, SMO2 and PASS with SN_0, SN_10 and SN_30 are laid out in tabular form on Tables 1, 2 and 3.

SN_0 means the predicted polyadenylation site is exactly the same as the EST-supported polyadenylation site. SN_10 means the EST-supported polyadenylation site is within 10 nucleotides of the predicted polyadenylation site. SN_30 means the EST-supported polyadenylation site is within 30 nucleotides of the predicted polyadenylation site. SP_CDS means the specificity achieved by running the classifier on coding sequences. SP_5UTR means the specificity achieved by running the classifier on 5'UTR sequences. SP_Intron means the specificity achieved by running the classifier on intronic sequences.

For Figure 5, the specificity of PASS 1.0 differs from what was reported in [3]. This is because the specificity calculated in [3] includes positions even where PASS 1.0 was unable to make a prediction as stated in Section 2.4. By doing so, it almost always recognizes those positions as TN, which will inevitably increase its specificity, whereas the calculation of specificity in this paper does not make use of those positions, hence resulting in a lower specificity.

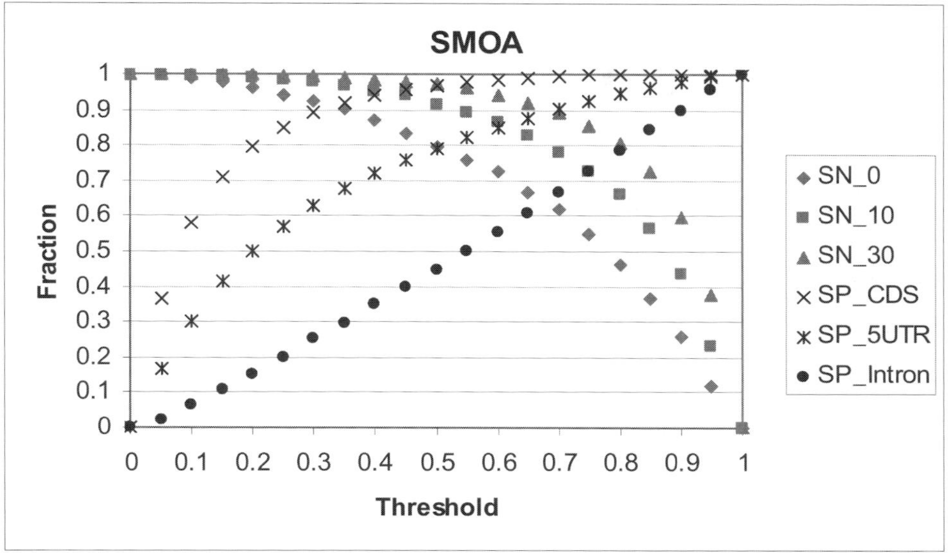

Figure 3. The prediction performance of SMOA.

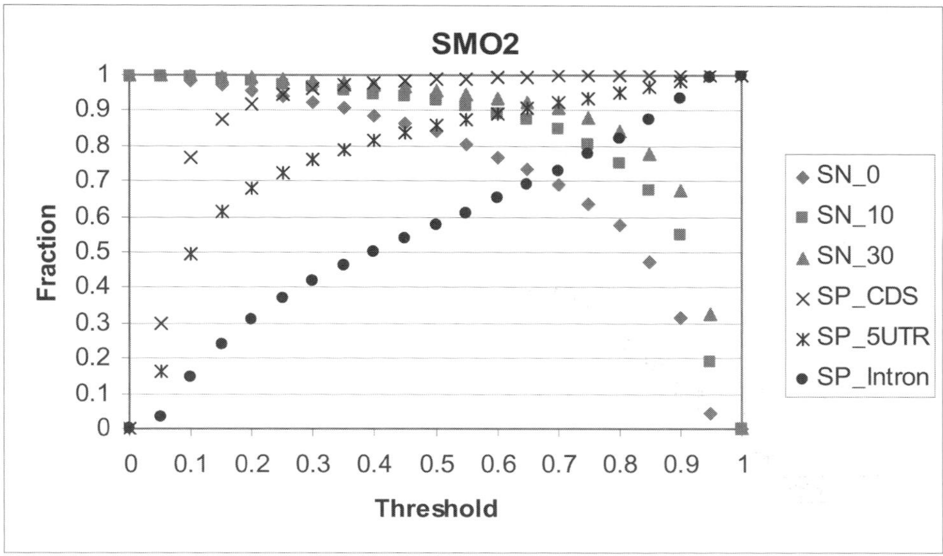

Figure 4. The prediction performance of SMO2.

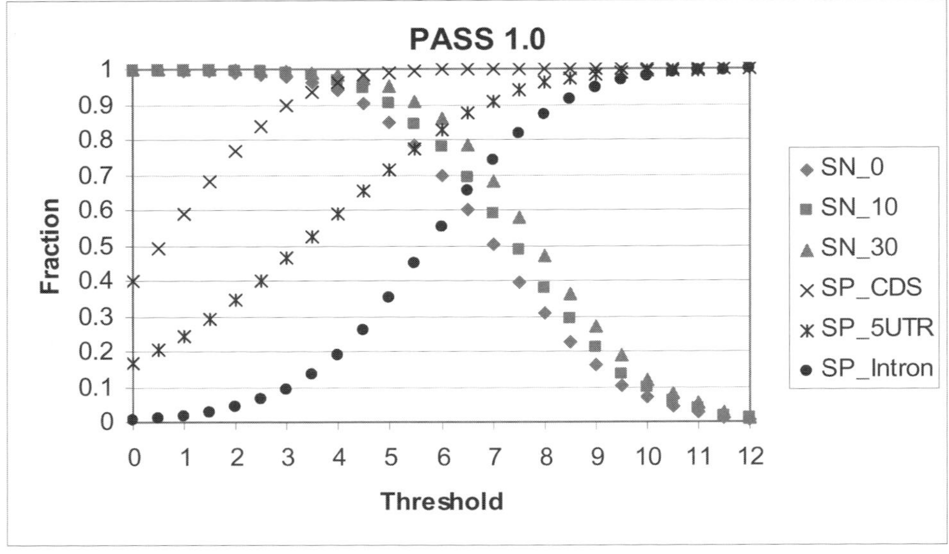

Figure 5. The prediction performance of PASS 1.0.

Table 1. Equal-error-rate points of SMO1, SMO2, and PASS 1.0 for SN_0.

SN_0	SMO A		SMO 2		PASS 1.0	
Control Sequences	SN & SP	Threshold	SN & SP	Threshold	SN & SP	Threshold
CDS	91.1%	0.33	94.3%	0.24	95.3%	3.76
5'UTR	79.3%	0.50	84.9%	0.48	77.7%	5.53
Intron	63.9%	0.68	71.1%	0.68	62.8%	6.36

Table 2. Equal-error-rate points of SMO1, SMO2, and PASS 1.0 for SN_10.

SN_10	SMO A		SMO 2		PASS 1.0	
Control Sequences	SN & SP	Threshold	SN & SP	Threshold	SN & SP	Threshold
CDS	94.8%	0.42	96.5%	0.31	96.5%	4.02
5'UTR	85.8%	0.61	89.2%	0.60	80.7%	5.81
Intron	72.5%	0.75	78.8%	0.76	67.7%	6.62

Table 3. Equal-error-rate points of SMO1, SMO2, and PASS 1.0 for SN_30.

SN_30	SMO A		SMO 2		PASS 1.0	
Control Sequences	SN & SP	Threshold	SN & SP	Threshold	SN & SP	Threshold
CDS	97.1%	0.50	97.5%	0.37	97.5%	4.29
5'UTR	89.8%	0.69	91.5%	0.67	84.0%	6.13
Intron	79.2%	0.81	83.0%	0.81	71.7%	6.85

4. Discussions

From the results, it is clear that introducing the cascade classifier (SMO2) step on top of the "feature generation, feature selection, feature integration" methodology helps improve both sensitivity and specificity (Shown by comparison between SMOA and SMO2). Our prediction model (SMO2) is also able to achieve a significant 7 – 11% higher sensitivity and specificity against PASS 1.0 with control sequences from 5'UTR and introns, while maintaining similar sensitivity and specificity with control sequences from the coding region.

When different control sequences are used, a drastic difference in the levels of sensitivity and precision is observed. With SN_0, coding sequences gives a sensitivity and specificity of 94.3% while introns sequences only achieved 71.1%; see Table 1. There could be two main reasons for this:

Firstly, recall that the parameter of our arabidopsis prediction model is set by using Dataset A derived by Hao Han. In that dataset, the control sequences used are from coding regions. This could have caused our arabidopsis prediction model to be biased towards coding sequences. Introducing 5'UTR and intronic sequences into the dataset used for setting the parameters could potentially increase the accuracy achieved on 5'UTR and intron.

Secondly, features used in our arabidopsis prediction model are compositional features. It is known that the intronic sequences have high A and U composition. This is a characteristic shared by the (+ve) sequences with EST supported polyadenylation site. This could explain why our arabidopsis prediction model has more difficulty in separating them. Therefore, introducing positional specific features or any features that are not compositional could potentially enhance the prediction model.

5. Conclusions

Our arabidopsis prediction model was built based on a machine learning methodology as described in [4]. It basically consists of 3 steps: 1) feature generation 2) feature selection 3) feature integration. An extra step (cascade classifiers) has been added for our model which helped increase sensitivity and specificity. In the above few steps, feature generation is the most crucial and difficult step. Given the "correct" set of features to generate, any method used for feature selection, feature integration and cascade classifier would yield high accuracy. However, to find the "correct" set of features to generate is similar to the problem of motif finding given a set of sequences, which is known to be NP-hard. Our approach is to use biological understanding of polyadenylation mechanism to decide on the set of features to generate. This approach has been shown to yield reasonably high accuracy in our prediction model.

Our arabidopsis prediction model has outperformed PASS 1.0 by a 7 – 11% margin on a validation dataset given by Qingshun Li (inventor of PASS 1.0). On the validation dataset, sensitivity and specificity for our model ranges from 71.1% to 94.3% with SN_0. As discussed previously, it is possible that the prediction model that we have developed for arabidopsis is biased towards coding sequences due to the parameter setting process. Therefore, one possible way to better improve its accuracy for other control sequences would be to redo the parameter setting procedure using other control sequences. Also, including more features that are not compositional could help better distinguish (+ve) sequences from (-ve) intronic sequences as they are compositionally similar.

As polyadenylation takes place just before the end of a transcription, the ability to accurately predict a polyadenylation site would be useful in predicting the ends of transcripts and also terminal exons. To this end, our prediction model has achieved reasonable accuracy and would certainly be useful for better gene annotations.

The datasets and source files used for this project are available at http://www.comp.nus.edu.sg/~wongls/projects/dnafeatures/giw07-supplement/

Acknowledgments

We thank Huiqing Liu and Hao Han for providing Dataset A and for their involvement in helpful discussions. We are also grateful to Qingshun Li for providing Dataset B, C, D and E.

References

[1] Emmanuel Beaudoing, Susan Freier, Jacqueline R. Wyatt, Jean-Michel Claverie and Daniel Gautheret, (2000). Patterns of Variant Polyadenylation Signal Usage in Human Genes. *Genome Research*, Vol. 10, No. 7, July, 2000, pp. 1001-1010.

[2] Diana F. Colgan and James L. Manley, (1997). Mechanism and regulation of mRNA polyadenylation. *Genes & Dev*, Vol. 11, No. 21, November 1, 1997, pp. 2755-2766.

[3] Guoli Ji, Jianti Zheng, Yingjia Shen, Xiaohui Wu, Ronghan Jiang, Yun Lin, Johnny C Loke, Kimberly M Davis, Greg J Reese and Qingshun Quinn Li, (2007). Predictive modeling of plant messenger RNA polyadenylation sites. *BMC Bioinformatics*, Vol. 8, No. 43, February 7, 2007.

[4] Huiqing Liu and Limsoon Wong, (2003). Data Mining Tools for Biological Sequences. *Journal of Bioinformatics and Computational Biology*, Vol. 1, No. 1, April 7, 2003, pp. 139-167.

[5] Johnny C. Loke, Eric A. Stahlberg, David G. Strenski, Brian J. Haas, Paul Chris Wood and Qingshun Quinn Li, (2005). Compilation of mRNA Polyadenylation Signals in Arabidopsis Revealed a New Signal Element and Potential Secondary Structures. *Plant Physiology*, Vol. 138, July 2005, pp. 1457-1468.

[6] J. Platt, (1999). Fast Training of support vector machines using sequential minimal optimization. B. Schölkopf, C. J. C. Burges, and A. J. Smola, editors, *Advances in Kernel Methods --- Support Vector Learning*, pages 185-208, Cambridge, MA, 1999. MIT Press.

[7] V.N. Vapnik, (1995). *The Nature of Statistical Learning Theory*. Springer-Verlag, Berlin, 1995.

[8] Ian H. Witten and Eibe Frank (2005). "*Data Mining: Practical machine learning tools and techniques*", 2nd Edition, Morgan Kaufmann, San Francisco, 2005.

COMPUTATIONAL ANALYSIS AND MODELING OF GENOME-SCALE AVIDITY DISTRIBUTION OF TRANSCRIPTION FACTOR BINDING SITES IN CHIP-PET EXPERIMENTS

VLADIMIR A. KUZNETSOV* [1]
kuznetsov@gis.a-star.edu.sg

YURIY L. ORLOV [1]
orlovy@gis.a-star.edu.sg

CHIA LIN WEI[1]
weicl@gis.a-star.edu.sg

YIJUN RUAN[1]
ruanyj@gis.a-star.edu.sg

*Correspondent author: V.A.K.
[1] Genome Institute of Singapore, Biopolis street, 60, 138672 Singapore

Advances in high-throughput technologies, such as ChIP-chip and ChIP-PET (Chromatin Immuno-Precipitation Paired-End diTag), and the availability of human and mouse genome sequences now allow us to identify transcription factor binding sites (TFBS) and analyze mechanisms of gene regulation on the level of the entire genome. Here, we have developed a computational approach which uses ChIP-PET data and statistical modeling to assess experimental noise and identify reliable TFBS for c-Myc, STAT1 and p53 transcription factors in the human genome. We propose a mixture probabilistic model and develop computational programs for Monte Carlo simulation of ChIP-PET data to define the background noise of the sequence clustering and to identify the probability function of specific DNA-protein binding in the eukaryotic genome. Our approach demonstrates high reproducibility of the method and not only distinguishes bona fide TFBSs from non-specific TFBSs with a high specificity, but also provides algorithmic and computational basis for further optimization of experimental parameters of the ChIP-PET method.

Keywords: ChIP-PET, transcription factor binding sites, human genome, mixture probabilistic model, Kolmogorov-Waring process, Monte Carlo simulation

1.   Introduction

Identification of gene regulatory elements for a given transcription factor is an important problem of computational genomics. The function of promoters, enhancers and other regulatory elements is mediated by DNA/protein interactions. The protein transcription factor binding sites (TFBS) serve as the basic units of gene functional activity. Computational prediction and high-throughput experimental validation of genome-scale sets of binding sites demands integrated approaches. Recently, great success has been achieved in the identification of TFBS for several essential regulators (p53 [9], c-Myc [2,11], STAT1 [3], p63 [10]) in human and Oct4, Sox2 and Nanog transcription factors (TFs) in mouse [6]. However, it has been difficult to identify all specific TFBS for several reasons. Currently available experimental information about the specificity of TF binding is essentially incomplete due to the difficulty of measuring the entire dynamical range of avidities of large (and actually unknown) numbers of DNA binding sites for a given TF and high level background noises vs. signals.

A recent development of sequencing-cloning technology [8] entails the possibility of highly efficient and unbiased coverage of mammalian genome for large-scale identification of regulatory elements (Chromatin ImmunoPrecipitation Paired-End diTag, or in brief, ChIP-PET method). ChIP-PET provides a new powerful technique for localization of the most physically specific mammalian TF binding regions at a resolution of up to a few base pairs [6,9,11]. The software suite for comprehensive processing and managing of raw Paired-End diTag (PET) sequence data were recently described in [1].

Most unexpectedly, all studies using ChIP-PET data have shown that the TFs bind specifically to a surprisingly large number of genomic regions (extrapolated to 5,000-20,000 depending on the protein) [6,9,10,11]. Due to a large data volume, the major fraction of these TFBSs would not be validated by traditional experimental methods. Our knowledge about optimization of the relationship between the specific and noise events are still limited. Therefore, new mathematical and computational models are required in order to analysis of raw ChIP-PET data and correctly identify and predict specific TF binding regions and to optimize parameters of ChIP-PET method.

In this work, we present a probabilistic model of protein-DNA binding and computational simulations that model the ChIP-PET experiment concerned with specificity and sensitivity issues of TFBSs detection. We study the performance of a new analytical approach using ChIP-PET data for human p53, c-Myc, IFN-α induced STAT1 and IFN-γ induced STAT1 [9,11]. Finally, we discuss some problems that arise with the avidity function of TFBS when applied on the scale of the entire genome and with the functionality of revealed TF binding sites.

2.   Data, Methods, Models, Algorithms and Software

2.1. *Transcription factors*

Transcription factor p53 regulates the expression of genes involved in a variety of cellular functions including cell cycle arrest, DNA damage repair, and apoptosis. ChIP-PET analysis of p53 binding in human colon cancer cells HCT116 was carried out as described in [9]. c-Myc is a proto-oncogene that regulates cell growth, cell proliferation, cell differentiation, and apoptosis [2]. In the ChIP-PET, we used human cell line that expresses high levels of exogenous c-Myc under the control of tetracycline [11]. STAT1 (signal transducer and activator of transcription) regulates proliferation by promoting growth arrest and apoptosis in response to interferon (IFN) signals [3]. ChIP-PET analysis of human cancer cells HelaS3 was carried out after treatment of these cells by IFN-α and INF-γ, as described in [3].

2.2. *Basic Concept of ChIP-PET Method*

Paired-End diTag (PET) method extracts a pair of 16-18 bp sequences from 5' end and 3' end of each cDNA clone, concatenates the PETs for efficient sequencing, and maps the resulting PET sequences to the genome. Such PET sequences characterize the ChIP enriched DNA fragments. Figure 1 shows a flow chart of ChIP-PET sequences processing, mapping and clustering to the genome (using c-Myc library obtained from human P493 cells as the example) [11].

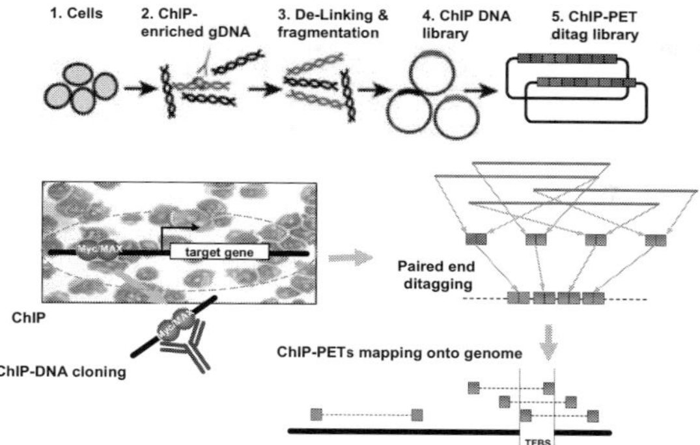

Figure 1. ChIP-PET. Top panel: Outline of the chromatin immunoprecipitation Paired-End diTag (ChIP-PET) method. Bottom panel: ChIP-PET analysis of c-Myc binding sites in human P493 cells. See details in [11]. Overlapping PET clusters define a more precise BS.

2.3. Definitions of DNA Fragment Cluster and Cluster Overlap

Let us define a *DNA fragment cluster* (or a cluster) as the overlapping PET DNA sequence fragments mapped to the genome (Figure 2A). More specifically, a PET sequence belongs to a cluster if it overlaps by at least 4 bp with any other sequence of the cluster in chromosome coordinates (Figure 2A). The number of PET sequences in a cluster is the *cluster size*. A total cluster span is defined as the genome region span covered by the cluster (Figure 2A). *The cluster overlap* is the most common PET DNA fragment in overlapping PETs in the given cluster. *The cluster member overlap count* (the peak) is the number of the overlapping PETs in a given cluster. The distribution of PET sequences within the cluster of sizes 3 and larger could be complex due to several cluster peaks (i.e., multimodal distribution of PET sequences). In this work, first, we count the highest peak (major mode) in the overlapping PET sequence cluster (Figure 2A). By examining the peaks observed for the rest of PET sequences in the cluster, we define the next highest peak and so on. To identify separate peaks in the given cluster, we use a strict definition of the cluster: every PET sequences in a cluster should overlap one another. To count the abundance of second peaks, we count the number of overlapped PET sequences excluding PET sequences from the first peak. If there are still sequences in the cluster, then we repeat the same procedure for third peaks and so on.

The number of cluster peaks occurrences is counted as the number of unique sequences containing at least one common nucleotide in a local PET sequence peak within a cluster. A cluster peak is more specific definition than a cluster overlap, because one cluster could contains more than one peak (local maximum of the sequence overlaps) and the peaks within a multimodal cluster could map the true protein-DNA interaction loci (Fig. 2A).

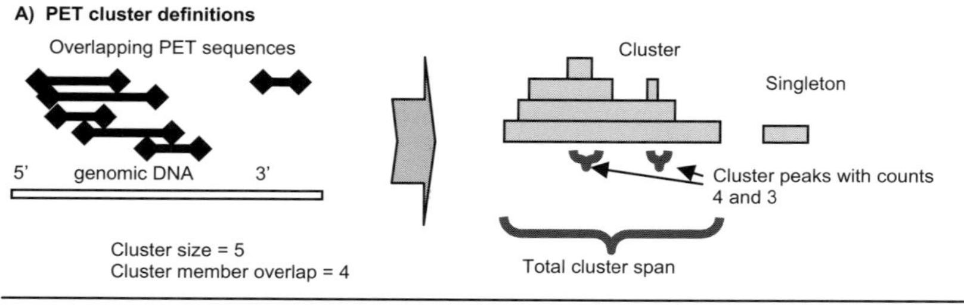

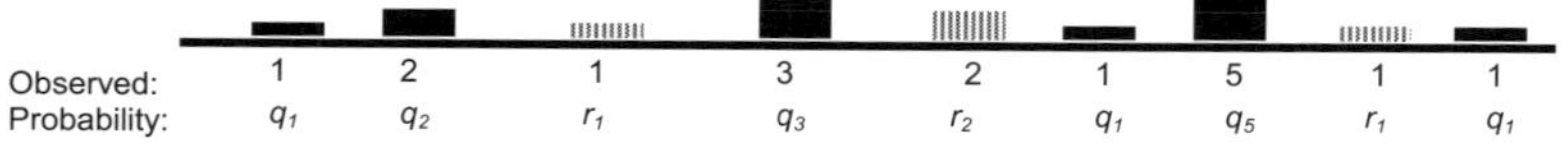

Figure 2. Two types of PET-clusters: definitions and statistical background of formation. A) Schematic example of sequence cluster, singleton, and cluster overlap. Cluster size is 5 (PET-5), and cluster member overlap is 4. Using strict criteria we define two clusters (peaks) by size 4 and 3. B) Schematic model of ChIP-PET sequences cluster overlaps on chromosome. q_1, q_2, q_3, q_5 are the binding probabilities for specific binding sites with avidity 1,2,3 and 5, respectively. r_1, r_2, r_3 are the probabilities of occurrence of sequence overlaps for non-specific PET sequences.

2.4. Characteristics of the ChIP-PET Libraries

In general, larger clusters (or peaks) often represent more specific binding sites [9,11] (see also Figure 2B). This correlation was observed for p53, ERE, c-Myc binding loci due to a concordance of the cluster (or picks) size with direct gene target expression data, direct qPCR-PET measurements and motif search analysis [9,11]. Nevertheless specific loci for the TFs could also be found in the smallest clusters (PET-2 peaks) and even in some singletons [11].

The number of PET sequences in the studied ChIP-PET libraries varied from 60 thousands (for p53 ChIP-PET library) to ~1 MB (for IFN-α activated STAT1 ChIP-PET library). The mean length of PET DNA sequences was in a range from ~400 bp to ~1700 bp (396 bp for c-Myc data; 623 bp for p53 data, 1385 bp for STAT1 data). Non-specific distinct PET DNA sequences represent a vast majority of PET sequences ranging from 75% to 95% of the total number of distinct sequences of ChIP-PET library.

2.5. Performance of ChIP-PET Data and Statistical Tasks

Significant amount of non-specific (background) genomic DNA is always present in the inmmunoprecipitated DNA material of ChIP-PET library. Some non-specific DNA might be easily filtered out after computer mapping of the DNA fragments on the genome [9]. Nevertheless background genomic DNA fragments that are uniquely mapped onto the genome still remain. With a larger sampling of DNA pool, the DNA fragments can be enriched by specific ChIP DNA sequences, and a larger number of true overlapping clusters might be observed. This sample size issue is related to the optimization of performance of the method. We have preliminary analyzed an influence of variation of several parameters of the method (e.g, frequency distribution of the lengths of PET sequences, derived after sonication and fragmentation of DNA-protein complexes, avidity of specific immuno-precipitation binding, etc.) on quality of ChIP-PET libraries. We have recognized that, *sampling* and *erroneous sequence* are essential issues in the analysis and validation of ChIP-PET data.

Due to background noise and sampling errors, the following basic statistical tasks are becoming imperative: i) to estimate specificity of the ChIP-PET experiment, i.e. to predict the number of reliable TFBSs; ii) to assign quantity measure of reliability to every PET cluster overlap peak that forms a putative TF binding site; iii) to predict the total number of specific binding sites presented in the PET library.

Using several data sets on PET sequence mapping onto human genome presented in T2G database [1], we analyze these problems via probabilistic modeling and computational simulation of non-specific and specific binding sites loci for a given TF.

2.6. Distributions of PET Cluster Overlaps and Clusters

The number of PET sequences covering specific genome sites should roughly relate to site avidity of binding protein (Figure 2B). We assume that the distribution function of distinct cluster size (observed by number of PET sequences in a peak) could be modeled as a sum of distributions of specific and non-specific (background noise) clusters (peaks):

$$P_{obs}(X{=}m) = \alpha * P_{sp}(X{=}m) + (1{-}\alpha) * P_{ns}(X{=}m), \qquad (1)$$

where P_{obs} is the probability distribution function of occurrence of a PET sequence cluster, X is the size of a given PET sequence cluster, $m{=}1,2,3,...$ is the number of sequences in a cluster, P_{sp} is the probability distribution function of specific PET cluster occurrence, $0{<}\alpha{<}1$ is the fraction of specific clusters in the cluster population, P_{ns} is the probability distribution function of occurrences of the non-specific (background noise) cluster in the cluster population.

Based on ChIP-PET data, we could construct an empirical frequency distribution function of occurrence of PET clusters and corresponding PET cluster peaks. P_{sp} is related to the specific avidity of DNA-protein binding. We can estimate P_{sp} using the Generalized Pareto probability function [4,5], which can be derived from the Kolmogorov-Waring distribution function as an asymptotic solution [5]. We could estimate P_{ns} by a computer simulation of the non-specific sequence clustering model. We will discuss this model in the next section.

2.7. A Data-driven Model of Background Noise Sequence Overlapping

To simulate non-specific component P_{ns}, we propose a physical model dependant on the chromosome size, in particular, virtual PET sequences of the observed length randomly drop down into an interval equal to the chromosome size. The algorithm is as follows: 1) Virtual random sequences mapped on a given chromosome randomly drop down into sequence domain that equals to available sequence domain of the chromosome; 2) Virtual position in the chromosome was selected by a random number generator; 3) The length of the virtual sequence is taken from the pool of observed PET sequences (we use empirical distribution of sequence lengths for our simulations (Figure 3)); 4) Virtual clusters are counted from overlapping virtual sequences (by at least 1 nt in the most common PET sequence overlap); 5) Y and M chromosomes and centromere regions of other chromosomes together PET sequences mapping these genome territories were excluded from modeling process.

The use of the observed length distribution is an important for modeling since longer sequences have a larger chance to form false clusters. An example of observed PET sequence length distribution can be found in Figure 3. It is possible to use a predefined fixed length for all virtual PET sequences (average length of observed PET sequences) [9]. We observed that such a model is oversimplified real data; in particular, it skips a number of large clusters (PET3+) which are important for unbias estimation of sensitivity of the method (Figure 3). Our data-driven noise model predicts a higher number of random clusters than the simplified model predicts.

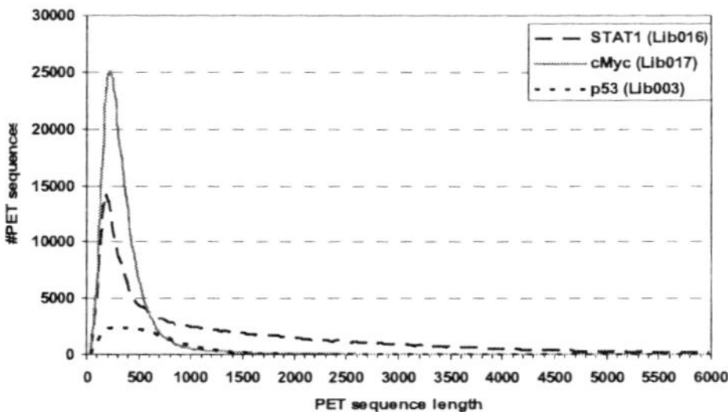

Figure 3. Observed PET sequence length distribution for p53, c-Myc and STAT1 libraries. The left size of the distributions could be fit well by the gamma distribution, however, the right long tail assumes more complex (mixture) distribution.

For the next step of estimation of model Eq (1), we performed a subtraction of the simulated distribution of random cluster size from the observed distribution of the cluster size. We estimated also a fraction $(1-\alpha)$ of non-specific sequences of model Eq (1) as a parameter to fit the observed frequency distribution of cluster size, assuming that frequencies of the non-specific clusters should be smaller than the one observed for each cluster size.

Notice. To overcome the problem of stochastic behavior, we fulfilled the simulation of random cluster formation many times (from 1 to 1000). Special attention was paid to quality of software for random number generation (RNG) [7].

2.8. A Model of Avidity Function of Specific Binding Site

We model the specific avidity distribution function using the truncated Generalized Discrete Pareto (GDP) function, which can be considered as a good limiting approximation of many random processes [4,5]:

$$f(m) := P_{sp}(X = m) = \zeta_J^{-1} \frac{1}{(m+b)^{k+1}}$$

$$(2)$$

where the $f(m)$ is the probability that a randomly chosen specific BS has an avidity value m. The f involves two unknown parameters, k, and b, where $k>0$, and $b>1$; the normalization factor ζ_J is the generalized Riemann zeta-function [5], truncated in the interval [2, 200]. Eq(2) can be considered as asymptotic distribution function derived from Kolmogorov-Waring (KW) probability function [5]. This KW model could be used as possible exploratory model of aggregation TF on a DNA binding site. In particular, we could model the evolution of TF-DNA interaction as the random the random linear Kolmogorov process [5] of binding and detachment of TF on specific DNA binding sites taken to account at least two binding transition probabilities: due to specific "binding potential" (preferential attachment mechanism [5]) and "non-specific potential" (Poisson process mechanism [5]). Similar two processes but with different intensities are assumed for detachments transitions.

2.9. *Analysis of Empirical Avidity Function of Specific Binding Sites*

Due to our findings, the shape of the avidity probability function of TF-DNA binding on the genome scale should be described as skewed function of avidity (Figure 4). An example of observed avidity function for c-Myc binding sites defined by ChIP-qPCR method is presented in Figure 4A. We have also found that the distribution of avidity measured by qPCR as well as the tail of the empirical probability functions of cluster overlaps (Figure 4A, Figure 5) follow the GDP and correlate to each other (Figure 4B). One can see skewed distribution approached by the power law.

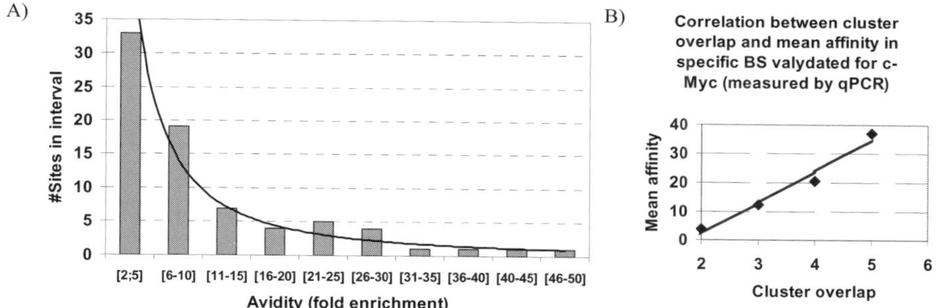

Figure 4. A) Avidity distribution function of PET sequences defined by ChIP-qPCR for c-Myc TFBS data. B) Correlation between ChIP-qPCR avidity and the number of sequences in specific cluster overlaps of c-Myc ChIP-PET library.

We argue that it is indeed a Pareto-like distribution. We have observed a similar avidity function for p53 [9] and Nanog transcription factor BS (not presented). Figure 4B shows a relation between the number of the sequences in cluster overlaps and the avidity value of c-Myc binding sites defined with ChIP-qPCR [11]. For 76 specific c-Myc binding sites in clusters, we found that the avidity of BS correlates with the number ChIP-PET sequences in cluster overlaps. The correlation coefficient between two data sets equals to 0.51 (p<0.01). Thus, we could use Eq. (2) as an empirical model of true avidity function of TF-DNA binding.

The avidity function of the ChIP-qPCR defining the specific TF binding sites can range from 2 to 200 fold enrichments [2]. Similar dynamical range was observed in our study of c-Myc mapping on the human genome [11].

Using this estimate, we calibrated enrichment fold from 2 to 200 and fitted the avidity function by the GDP function using the method presented in [5]. We constructed avidity function distribution for proposed number of binding sites that best fit the observed data based on minimal assumptions of statistical parameters of the distribution (data are not shown). For example, parameters k and b for $f(m)$ function from Eq. (2) for c-Myc binding sites are equal to 3.4 and 1, correspondingly.

3.   Results of Numerical Modeling

3.1. *Goodness of Fit Analysis of Mixture Model and Estimations*

To analyze the observed PET cluster peaks distribution, we parameterized the specific (Eq. 2) and nonspecific probability functions in Eq. (1) and estimated the parameter α. To accomplish that, we first used a goodness of fit analysis method presented below. Figure 5 shows an example of our tail-fitting and extrapolation method. We fit the theoretical (power law) function using only right (specific) part of the cluster size distribution (Figure 5A), and fit the proposed exponential function using only left (non-specific) part of the same distribution (Figure 5B).

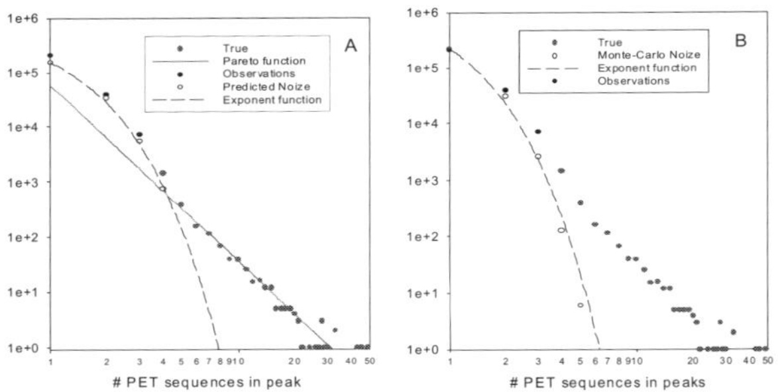

Figure 5. Decomposition of observed frequency distribution of the number of PET sequences in peaks in the STAT1 IFN-gamma activated library using (A) tail-fitting and extrapolation method and (B) noise peaks simulation and subtraction method. Red point: observed data; empty circles: restored background noise data; black circles data for the noise (nonspecific) peaks. Solid line on panel A: best-fit power law distribution with exponent parameter k=2.2 +/- 0.82 (p<0.01). Dashed lines: exponential function with slope parameter 1.75+/-0.144 (p=0.02) and 2.32+/-0.010 (p<0.01) on panel A and panel B, respectively.

This figure shows the decomposition of the observed frequency distribution of the number of PET cluster peaks in the PET library for IFN-γ activated STAT1 (Lib016). To fit the parameters, we used Sigma-Plot software. In our model (Eq.1), best-fit power law distribution (Eq. 2) has exponent parameter k=2.2; the exponential function having a slope parameter 1.75 fits well to the non-specific (noise) component of the Eq.1. We assumed that all the cluster peaks of size greater than the cut-off value (3 or 4) are specific. By counting the total number of PET sequences associated with entire best-fit power distribution and the total number of PET sequences associated with peaks of the observed distribution, we can also estimate the fraction of specific PET sequences in the library, α (Eq.1). For instance, for STAT1 data, α equals to 0.26 (Table 1).

For STAT1 IFN-γ activated ChIP-PET library, we used goodness of fit analysis [5](Figure 5) and computer simulations (Figure 6). Starting with a simulation model of

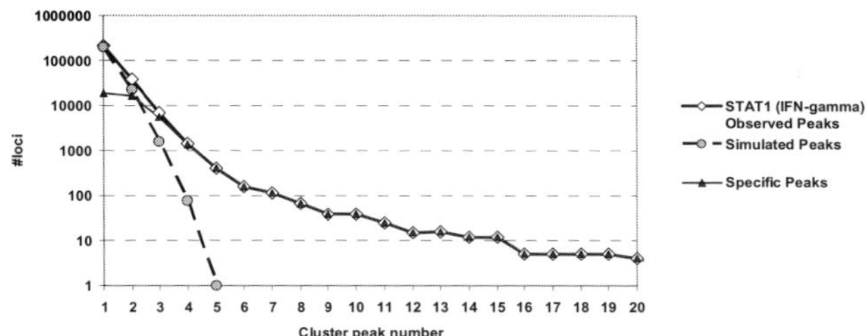

Figure 6. Observed and simulated noise distributions of cluster peak size for STAT1 library (Lib016). Specific cluster peak size distribution is estimated by subtracting the averaged simulated number of cluster peaks from observed number of cluster peaks. The noise distribution function was estimated by averaging of 20 Monte Carlo simulations. The fraction of noise PET sequences in our simulation was 75% of the STAT1 library size.

background noise distribution function, we can carry out a decomposition of the specific and non-specific components in Eq. (1). Figure 6 presents the noise peaks simulation and illustrates our subtraction method. To do that, we can calculate the distribution of non-specific (noise) cluster peaks and then after subtracting this distribution from observed distribution we could reconstruct the specific probability distribution function (Figure 6).

The both our fitting methods provide similar frequency distributions of the number of nonspecific (noise) PET sequence cluster peaks. Given the simulated and observed frequency distributions of the number of specific PET sequences within cluster peaks and subtracting one distribution from another, we can derive the true frequency distribution of the number of specific peaks, reporting here as the true TFBS loci (Figure 6). Thus, our results support the suggestion that the probability distribution of an avidity of specific binging sites follows a Pareto-like law while the non-specific components would be well approximated by exponential function.

We constructed also truncated cumulative function of the number of specific clusters (sum of number of clusters of size larger than given count). We used this cumulative function to estimate a fraction of specific TFBS loci for a given PET peak (Table 1). Due to estimated background cut-off value, this fraction equals to 1.0 for STAT1 PET6+ clusters (Table 1), that means that a fraction of non-specific cluster peaks larger than six is a negligibly small.

3.2. *Specificity of PET Clusters*

Based on our models, we estimate a cut-off critical value of cluster size which discriminates non-specific binding sites (noise) from specific binding sites in a given ChIP-PET library. The degree of reliability of every cluster in the library could be assigned

Table 1. Observed and simulated cluster peak size distribution for
STAT1 IFN-γ activated library (Lib016).

PET count	1	2	3	4	5	6	7	8	9+
#Observed peaks	212447	39025	7103	1444	400	157	113	66	198
#Non-specific peaks	193528	22432	1565	75	1	0	0	0	0
Cumulative # observed peaks	260953	48506	9481	2378	934	534	377	264	198
Cumulative # specific peaks	43352	24433	7840	2302	933	534	377	264	198
Cumulative specific fraction	0.166	0.504	0.827	0.968	0.999	1	1	1	1

Table 2. Cut-offs of PET cluster size (95% level) and number of
putative specific BS estimated by observed distributions.

TF	Total % of spec. PET	Cut-off value	% Spec. at cut-off	#BS at cut-off	% Spec. in PET2+	#BS in PET2+	#BS in PET1+
p53	5	3	98.6	284	53.2	936	1659
STAT1(IFN-γ)	25	4	96.8	2302	50.3	24433	43352
STAT1(control)	13	5	96.2	52	32.5	11735	18743
c-Myc	5	4	93.6	44	28.1	3683	9295

based on the probability that a cluster of the given size contains a binding site. Thus we can annotate every region in the clusters either as reliable (the probability that it contains BS is more than 99%), or probable (probability >95%), or potential (>70%).

We simulated background noise distribution by varying the number of specific PET sequences in the experiment from 0% to 30% and by selecting the best fit parameters. For p53 data we estimate the total percent of specific PET sequences in the ChIP-PET library to be at 5% (Table 2, second column). Comparing simulated and observed distributions of PET sequence cluster peak size, we can estimate the number of specific BS at any fixed cut-off value. For p53 data, for example, the cut-off value for cluster peak size equals 3 (third column) while specificity at this cut-off (number of specific BS in PET3+ cluster peaks) equals 284 (Table 2, fourth column). The cut off value of cluster peak size defines the specificity to be greater than 95%. The 6-th and 7-th columns in Table 2 demonstrate the specificity in PET2+ cluster peaks. The specificity is much lower for p53 BSs (53.2%), but the estimated total number of BSs in PET2+ cluster peaks is much higher (936 BSs for p53, Table 2). The total number of putative BS predicted by the PET singletons and clusters is estimated to be 1659. This is an estimate of the low limit of the total number of specific p53 BSs in the human genome. Our extrapolation of the GDP model predicts >5000 p53 BSs, most of which, however, should be very low-avidity BSs and therefore should be not functional in a cell.

The cut-off values of cluster peak size are relatively large for STAT1 and c-Myc libraries (Table 2). The total number of specific binding sites in the human genome is estimated to be as large as up to 43000 for IFN-γ induced STAT1 TF. By our computational simulations, the smallest number of specific BS estimated represented by the c-Myc library is ~ 9000. Base on extrapolation of the best-fit model (1), however, the total number of specific c-Myc BSs in the human genome should about 20000 or even larger [11].

4. Discussion and Conclusion

We have developed a computational method to estimate the number of specific binding loci in the genome for transcription factors studied in ChIP-PET experiments. Our probabilistic model of ChIP-PET clusters permits an accurate estimation of parameters of TFBS avidity function based on ChIP-PET experiment. The model explicitly uses information about the length of the ChIP-PET DNA fragments, the number of PET sequences in the library and the chromosome length. The summary is as follows:

1. We developed a statistical method for selecting specific **a** component in observed PET cluster sites distribution based on a novel mixture distribution model.
2. We developed a Monte Carlo simulation model and a program for non-specific binding events (background noise) in ChIP-PET cluster size distribution.
3. This computational model provides cut-off values for specific TFBS and supports estimates obtained by the goodness of fit method [9,11].
4. We have shown that the true and noise probability distributions of loci avidity are scale-dependent skewed functions: when the library size and/or the average sequence span become larger, the shape of each distribution changes. In particular, the tail of the distributions of occurrence of true BSs becomes longer.
5. The probability distribution of specific avidity of DNA-protein interactions for different TFs in mammalian cells can be described by a generalized Pareto function, while the random probability distribution is fitted by the exponential function.

Significant correlation between ChIP-qPCR avidity and the number of sequences in specific cluster overlaps of ChIP-PET data suggests that the size of specific PET sequence clusters could indeed reflect of DNA-TF avidity. Validation of predicted "true" BS was shown for p53 and c-Myc binding sites in PET3+ clusters and verified by qPCR experiments [9,11]. We note that the binding specificity of TF sites defined by our model does not mean that the functionality of the BS in a given cell. The DNA-TF avidity is really very complex and skewed function (Eq (2), Figure 4) of many factors. Other factors such as the state of chromatin remodeling, epigenetic factors (acetylation and methylation of histones and DNA) may affect both ChIP-PET and ChIP-qPCR outcomes. Recently, Abcam ChIP Grade antibodies in combination with the new Solexa 1G sequencing technology have been used to identify the patterns of histone methylations on the human genome scale [13]. We assume that the computational modeling of binding avidity of specific TF BSs and mapping histone methylation regions could provide better understanding of TF binding and epigenetic control of DNA-TF avidity. It is interesting to note that recent ChIP-seq (ChIP-sequencing) methods have revealed 41582 putative STAT1-binding regions in IFN-γ stimulated HeLa S3 cells [12]. However, avidity of the vast majority of these BSs due to our model should be very low and perhaps these low-avidity BSs are unlikely functional. The addition of sequence information (consensus or weight matrix for given protein binding site) and genome information (as proximity to promoter regions, CpG islands) could significantly improve the quality of estimations for the binding sites [9,11,12]. ChIP-PET technology as well as ChIP-on-chip, ChIP-seq and STAGE (Sequence Tag Analysis of Genomic Enrichment) technologies [3,12] in combination with simulation analysis will allow us to identify thousands of novel binding sites in the human genome. A challenging statistical problem of estimation of specificity and sensitivity of these methods could be solved using the approach suggested in this work.

Acknowledgments

The authors would like to acknowledge Dr. Chiu Kuo Ping for useful discussions. This work is supported by the A*STAR of Singapore.

References

[1] Chiu, K.P., Wong, C.H., Chen, Q., et al, PET-Tool: a software suite for comprehensive processing and managing of Paired-End diTag (PET) sequence data, *BMC Bioinformatics*, 7:390, 2006.

[2] Fernandez, P.C., Frank, S.R., Wang, L., Schroeder, M., Liu, S., Greene, J., Cocito, A., and Amati, B., Genomic targets of the human c-Myc protein, *Genes Dev.*, 17(9):1115-29, 2006.

[3] Hartman, S.E., Bertone, P., Nath, A.K., et al. Global changes in STAT target selection and transcription regulation upon interferon treatments, *Genes Dev.*, 19(24):2953-68, 2005.

[4] Johnson, N.L., Kotz, S., and Balakrishnan, N., *Discrete Multivariate Distributions*, John Wiley&Sons, Inc., New York 299 p., 1997.

[5] Kuznetsov, V.A., Family of skewed distributions associated with the gene expression and proteome evolution, *Signal Processing*, 83:889-910, 2003.

[6] Loh, Y.H., Wu, Q., Chew, J.L., et al. The Oct4 and Nanog transcription network regulates pluripotency in mouse embryonic stem cells, *Nat Genet.*, 38(4):431-440, 2006.

[7] Matsumoto, M., and Nishimura, T., Mersenne Twister: A 623-dimensionally equidistributed uniform pseudo-random number generator, *ACM Transactions on Modeling and Computer Simulation*, 8:3-30, 1998.

[8] Ng, P., Wei, C.-L., Sung, W.K., et al., Gene identification signature (GIS) analysis for transcriptome characterization and genome annotation, *Nat Methods*, 2:105-111, 2005.

[9] Wei, C.L., Wu, Q., Vega, V.B. et al., A global map of p53 transcription-factor binding sites in the human genome, *Cell*, 124(1): 207-19, 2006.

[10] Yang, A., Zhu, Z., Kapranov, P., McKeon, F., Church, G.M., Gingeras, T.R., and Struhl, K., Relationships between p63 binding, DNA sequence, transcription activity, and biological function in human cells, *Molecular Cell*, 24:593-602, 2006.

[11] Zeller, K.I., Zhao, X., Lee, C.W., et al. Global mapping of c-Myc binding sites and target gene networks in human B cells, *Proc Natl Acad Sci U S A*, 103(47):17834-17839, 2006.

[12] Robertson, G., Hirst, M., Bainbridge, M. et al., Genome-wide profiles of STAT1 DNA association using chromatin immunoprecipitation and massively parallel sequencing, *Nat Methods.* 4(8):651-7, 2007.

[13] Barski A., Cuddapah S., Cui K. et al., High-resolution profiling of histone methylations in the human genome. Cell. 129(4):823-37, 2007.

Linear-Time Reconstruction of Zero-Recombinant Mendelian Inheritance on Pedigrees without Mating Loops

LAN LIU[1] TAO JIANG[1]

lliu@cs.ucr.edu jiang@cs.ucr.edu

[1] *Department of Computer Science and Engineering, University of California, Riverside, CA, USA*

With the launch of the international HapMap project, the haplotype inference problem has attracted a great deal of attention in the computational biology community recently. In this paper, we study the question of how to efficiently infer haplotypes from genotypes of individuals related by a pedigree *without mating loops*, assuming that the hereditary process was free of mutations (*i.e.* the Mendelian law of inheritance) and recombinants. We model the haplotype inference problem as a system of linear equations as in [10] and present an (optimal) linear-time (*i.e.* $O(mn)$ time) algorithm to generate a *particular solution* * to the haplotype inference problem, where m is the number of loci (or markers) in a genotype and n is the number of individuals in the pedigree. Moreover, the algorithm also provides a *general solution* † in $O(mn^2)$ time, which is optimal because the size of a general solution could be as large as $\Theta(mn^2)$. The key ingredients of our construction are (i) a fast consistency checking procedure for the system of linear equations introduced in [10] based on a careful investigation of the relationship between the equations (ii) a novel linear-time method for solving linear equations without invoking the Gaussian elimination method. Although such a fast method for solving equations is not known for general systems of linear equations, we take advantage of the underlying loop-free pedigree graph and some special properties of the linear equations.

Keywords: haplotype inference, pedigree analysis, mating loop, tree-pedigree, linear-time algorithm, system of linear equation, general solution.

1. Introduction

In October 2002, a multi-country collaboration, namely, the international *HapMap* project was launched [5]. One of the main objectives of the HapMap project is to identify the haplotype structure of humans and common haplotypes among various populations. This information will greatly facilitate the mapping of many important disease-susceptible genes. However, due to the diploid structure of the human genome, in practice, genotype data, instead of haplotype data are collected routinely, especially in large-scale sequencing projects mainly to save experimental costs. Hence, combinatorial algorithms as well as statistical methods for the inference of haplotypes from genotypes, which is also commonly referred to as *phasing*, are urgently needed and have been intensively studied in the literature.

This paper is concerned with the inference of haplotypes from genotypes of individuals related by a *pedigree*, which describes the parent-offspring relation-

---

*A particular solution of any linear system is an assignment of numerical values to the variables in the system which satisfies the equations in the system.

†A general solution of any linear system is denoted by the span of a basis in the solution space to its associated homogeneous system, offset from the origin by a vector, namely by any particular solution. A general solution for ZRHC is very useful in practice because it allows the end user to efficiently enumerate all solutions for ZRHC and performs tasks such as random sampling.

ship among the individuals. Two biological assumptions will be made, namely, the *Mendelian law of inheritance, i.e.* one haplotype of each child is inherited from the father while the other is inherited from the mother free of mutations, and the *zero-recombinant principle* which says that genetic recombination is very rare or nonexistent for closely linked markers and thus in haplotype inference we prefer haplotype configurations with zero recombinants if they exist [3, 9]. The problem of inferring zero-recombinant haplotypes from given genotypes under the Mendelian law, where we would like to enumerate all haplotype solutions that require no recombinants (if such solutions exist), is called the *zero-recombinant haplotype configuration (ZRHC)* problem and was studied recently in [6, 10]. When the input pedigree is a tree pedigree with no mating loops (which is often true for human pedigrees), the ZRHC problem is called the *Loop-Free ZRHC* problem. Figure 1 gives an illustrative example of pedigree, genotype, haplotype, as well as *recombination* where the haplotypes of a parent recombine to produce a haplotype of her child. (See the appendix for a more formal definition of the above biological concepts and computational problems).

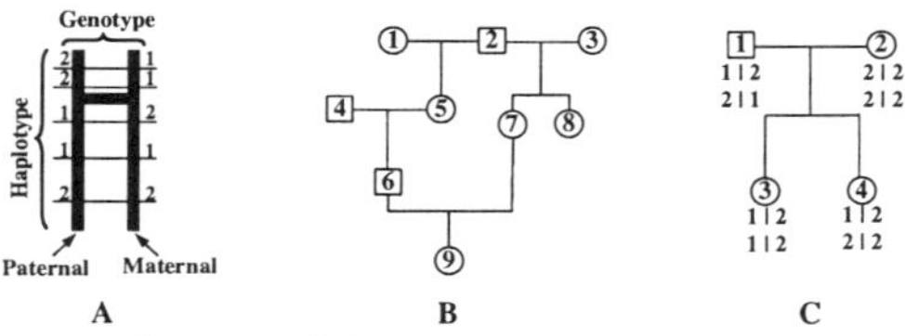

Fig. 1.   (A). The structure of a pair of chromosomes from a mathematical point of view. In the figure, each numeric value (1 or 2) represents a marker state or an *allele*. The haplotype inherited from the father (or the mother) is called *paternal* haplotype (or *maternal* haplotype, respectively). The paternal and maternal haplotypes are thus strings 22112 and 11212, and they form the genotype $\{1,2\}\{1,2\}\{1,2\}\{1,1\}\{2,2\}$, which is a string of unordered pairs of alleles at each locus.   (B). An illustration of a pedigree with 9 members with a mating loop, where circles represent females and boxes represent males. Children are shown under their parents with line connections. For example, individuals 7 and 8 are children of individuals 2 and 3. Individuals without parents, such as individuals 1 and 2 are called *founders*. A pedigree without mating loops is called a *tree pedigree* conventionally.   (C). An example of recombination event where the haplotypes of individual 1 recombine to produce the paternal haplotype of individual 3. The numbers inside the circles/boxes are individual IDs. Here, a "|" is used to indicate the phase of the two alleles at a marker locus, with the left allele being paternal and the right maternal. Both loci of individual 2 and the 2nd locus of individual 4 are *homozygous* while all the other loci in the pedigree are *heterozygous*.

The Algorithmic Problem, Previous Results and Our Result. The ZRHC problem can also be stated abstractly as a simple inheritance reconstruction problem as follows [10]. We have a pedigree connecting n individuals where each individual j has a genotype given in the form of string $\{a_{j,1}, b_{j,1}\} \cdots \{a_{j,m}, b_{j,m}\}$ defined on m marker loci. We would like to reconstruct two haplotypes (*i.e.* strings) $a_{j,1} \cdots a_{j,m}$ and $b_{j,1} \cdots b_{j,m}$ for each individual j such that the inferred haplotypes obey the Mendelian law and the zero-recombinant principle. That is, suppose that haplotypes $a_{j,1} \cdots a_{j,m}$ and $b_{j,1} \cdots b_{j,m}$ are inherited from j's father j_1 and mother j_2 respectively, then we have $a_{j,1} \cdots a_{j,m} \in \{a_{j_1,1} \cdots a_{j_1,m},\ b_{j_1,1} \cdots b_{j_1,m}\}$ and $b_{j,1} \cdots b_{j,m} \in \{a_{j_2,1} \cdots a_{j_2,m},\ b_{j_2,1} \cdots b_{j_2,m}\}$.

Li and Jiang presented an $O(m^3 n^3)$ time algorithm for ZRHC by formulating it as a system of $O(mn)$ linear equations with mn variables over the finite field $F(2)$ and applying Gaussian elimination [6]. Although this cubic time algorithm is reasonably fast, it is inadequate for large scale pedigree analysis where both m and n can be in the order of tens or even hundreds, and we may have to examine many pedigrees and haplotype blocks. There are, for example, over five million SNP markers in the public database dbSNP [5]. Recently, Xiao *et al.* [10] presented a much faster algorithm for ZRHC with running time $O(mn^2 + n^3 \log^2 n \log \log n)$ by introducing a new system of $O(mn)$ linear equations over $F(2)$ with mn variables and reducing it to an effectively equivalent system with only $O(n \log^2 n \log \log n)$ equations and at most $2n$ variables.

For Loop-Free ZRHC, the method proposed by Xiao *et al.* [10] runs in time $O(mn^2 + n^3)$ to produce a general solution and in time $O(mn + n^3)$ to produce a particular solution. Chan *et al.* [4] proposed a linear-time (*i.e.* $O(mn)$ time) algorithm to find a particular solution. It is unclear how the algorithm can be extended to produce general solutions efficiently. Moreover, the algorithm has four different stages and needs to maintain various consistency conditions (called SNP-consistency, Mendelian consistency, and endgame-consistency) and avoid some inconsistency conditions (such as the family problem) at these stages (see [4] for the details of the (in)consistency conditions). As a result, the analysis of the algorithm (concerning both correctness and complexity) is quite involved.

In this paper, we present a linear-time algorithm to generate a particular solution for Loop-Free ZRHC. Moreover, the algorithm can also provide a general solution in $O(mn^2)$ time, which is optimal because the size of a general solution could be as large as $\Theta(mn^2)$. [a] Our construction is based on the system of linear equations/constraints introduced in [10], a careful examination of the relationship between the constraints using the so called *constraint graphs* (to be defined later on), and a novel linear-time method for solving linear systems without invoking Gaussian elimination. In the algorithm, we first build a *one-to-one* mapping from the constraints to the edges in the pedigree graph, then solve the constraints by a single BFS traversal on the pedigree graph. We also give a rigorous proof of the correctness and tight analysis of the time complexity of our algorithm.

The rest of the paper is organized as follows. In section 2, we describe a system of linear equations for ZRHC and some useful graphs derived from the input pedigree introduced in [10] to make the paper self-contained. A linear-time algorithm to check if the linear equations derived from an input tree pedigree are consistent (*i.e.* if solutions for the linear system and thus ZRHC exist) and solve the linear system to attain a general solution is presented in section 3. The method of obtaining a particular solution by the algorithm is also described in this section. Some concluding remarks are given in section 4. Due to page limit, all proofs, some detailed biological definitions and an example execution of the main algorithm will be shown in the full version [1].

2. A System of Linear Equations and Equivalent Linear Constraints

In this section, we present a formulation of ZRHC in terms of linear equations and define some graph structures that will be used in our algorithm. The linear system and definitions were all introduced in [10], and are included here for the convenience of the reader.

2.1. *The Linear System*

Throughout this paper, n denotes the number of the individuals (or members) in the input pedigree and m denotes the number of marker loci. Without loss of generality,

---

[a]An instance of the Loop-Free ZRHC problem that has a general solution of size $\Theta(mn^2)$ is given in the appendix.

suppose that each allele in the given genotypes is numbered numerically as 1 or 2 (*i.e.* the markers are assumed to be *bi-allelic*, which make the hardest case for ZRHC [6, 10]), and the pedigree is free of genotype errors (*i.e.* the two alleles at each locus of a child can always be obtained from her respective parents). Hence, we can represent the genotype of member j as a ternary vector $\mathbf{g}_j$ as follows: $g_j[i] = 0$ if locus i of member j is homozygous with both alleles being 1's, $g_j[i] = 1$ if the locus is homozygous with both alleles being 2's, and $g_j[i] = 2$ otherwise (*i.e.* the locus is heterozygous). For any heterozygous locus i of member j, we use a binary variable $p_j[i]$ to denote the *phase* at the locus as follows: $p_j[i] = 0$ if allele 2 is paternal, and $p_j[i] = 1$ otherwise. When the locus is homozygous, the variable is set to $g_j[i]$ for some technical reasons (so that the equations below involving $p_j[i]$ will hold). Hence, the vector $\mathbf{p}_j$ describes the paternal and maternal haplotypes of member j.

Observe that the vectors $\mathbf{p}_1, \ldots, \mathbf{p}_n$ represent a complete haplotype configuration of the pedigree. In fact, the sparse linear system in [6] was based on these vectors. Also for technical reasons, define a vector $\mathbf{w}_j$ for member j such that $w_j[i] = 0$ if its i-th locus is homozygous and $w_j[i] = 1$ otherwise. Suppose that j is a non-founder member with her father and mother being j_1 and j_2, respectively. We define $\mathbf{d}_{j_r,j}(r = 1, 2)$ as follows: $\mathbf{d}_{j_1,j}$ be the vector $\mathbf{0}$ and $\mathbf{d}_{j_2,j} = \mathbf{w}_j$.

Utilizing the above definitions, we can formally express the ZRHC problem as a system of linear equations (refer to [10] for the details of deriving the linear system) over $F(2)$:

$$\begin{cases} p_k[i] + h_{k,j} \cdot w_k[i] = p_j[i] + d_{k,j}[i] & 1 \leq i \leq m,\ 1 \leq j, k \leq n, k \text{ is a parent of } j \\ p_j[i] = g_j[i] & 1 \leq i \leq m,\ 1 \leq j \leq n,\ g_j[i] \neq 2 \\ w_j[i] = 1 & 1 \leq i \leq m,\ 1 \leq j \leq n,\ g_j[i] = 2 \\ w_j[i] = 0 & 1 \leq i \leq m,\ 1 \leq j \leq n,\ g_j[i] \neq 2 \\ d_{k,j}[i] = w_j[i] & 1 \leq i \leq m,\ 1 \leq j, k \leq n,\ k \text{ is the mother of } j \\ d_{k,j}[i] = 0 & 1 \leq i \leq m,\ 1 \leq j, k \leq n,\ k \text{ is the father of } j \end{cases}$$

$$(1)$$

where $g_j[i], w_j[i], d_{k,j}[i]$ are all constants depending on the input genotypes, and $p_j[i], h_{k,j}$ are the unknowns. Note that, the number of p-variables is exactly mn and the number of h-variables is at most $2n$ since every child has two parents and there are at most n children in the pedigree.

Remark: Observe that for any member j, if the member itself or one of its parents are homozygous at locus i, $p_j[i]$ is fixed based on Equation (1). In the rest of the paper, we will assume that all such variables $p_j[i]$ are *pre-determined* (without any conflict), and use them as "anchor points" to define some new constraints about the h-variables.

2.2. *The Pedigree Graph and Locus Graphs*

To conform with standard graph theory notations, we transform the input pedigree into a graph, called the *pedigree graph*, by connecting each parent directly to her children, as shown in Figure 2 (B). Although the edges in the pedigree represent the inheritance relationship between a parent and a child and are directed, we will think of the pedigree graph, and more importantly, the subsequent *locus graphs* (to be defined below), as undirected in future definitions and constructions, since each edge (j, k) of the pedigree graph (and locus graphs) will be used to represent the constraint between the vectors $\mathbf{p}_j$ and $\mathbf{p}_k$ (*i.e.* the phases at j and k) via the variable $h_{j,k}$, which is symmetric. [b]

---

[b]The reader can also verify that the direction of an edge will not affect the graph traversal and the ensuing treatment of constraint equations to be discussed in the next two sections.

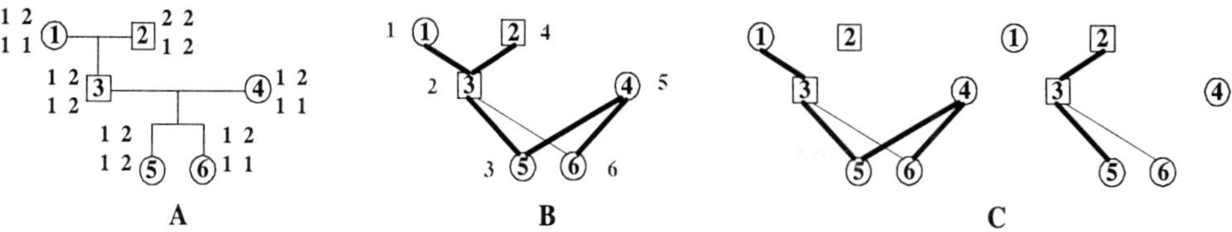

Fig. 2.   (A). An example pedigree with genotype data. Here, the alleles at a locus are ordered according to their id numbers instead of phase (which is unknown).   (B). The pedigree graph with a spanning tree. The tree edges are highlighted. Observe that there lies a *cycle* of length 4 in the given tree pedigree graph. The root of the tree is v_1. The number (in blue) next to a vertex is the index generated by a BFS traversal of the spanning tree. For instance, v_3 is v_2's predecessor in the BFS traversak where $index(v_3) = 2$ and $index(v_2) = 4$, while v_2 is v_3's biological father.   (C). The locus graphs. The left graph is for the first locus, which has a cycle, while the right graph is for the second locus. The locus forests are highlighted.

Clearly, such a pedigree graph $G = (V, E)$ may be cyclic due to mating loops or multiple children shared by a pair of parents. Let $\mathcal{T}(G)$ be any spanning tree of G. We also set an arbitrary vertex as the *root* of $\mathcal{T}(G)$. We start a BFS traversal on $\mathcal{T}(G)$ from the root, and index the vertices in E based on the order of their first-time occurrences in the traverse. For a vertex j, we denote by $index(j)$ the index of j. For instance, $index(r) = 1$ if vertex r is the root of $\mathcal{T}(G)$. If the last edge on the path from the root to vertex j is (k, j), k is called the *predecessor* of j, and denoted by $pred(j)$. [c] Obviously, $index(pred(k)) < index(j)$ for every non-root vertex j. $\mathcal{T}(G)$ partitions the edge set E into two subsets: *tree edges* and *non-tree edges*. For simplicity, the non-tree edges will be called *cross edges*. Let E^x denote the set of cross edges. Since $|E| \le 2n$ and the number of edges in $\mathcal{T}(G)$ is $n - 1$, we have $|E^x| \le n + 1$. Figure 2 (B) gives an example of tree edges, cross edges and an indexing on vertices.

For any fixed locus i, the value $w_k[i]$ can be viewed as the *label* of each edge $(k, j) \in E$, where k is a parent of j. We construct the *i-th locus graph* G_i as the subgraph of G induced by the edges with label 1. Formally, $G_i = (V, E_i)$, where $E_i = \{(k, j) | k \text{ is a parent of } j, w_k[i] = 1\}$. The i-th locus graph G_i induces a subgraph of the spanning tree $\mathcal{T}(G)$. Since the subgraph is a spanning forest, it will be referred to as the *i-th locus forest* and denoted by $\mathcal{T}(G_i)$. Figure 2 (C) shows the locus graphs and the locus forests of the given pedigree.

The locus graphs can be used to identify some implicit constraints on the h-variables as follows. First, we need "symmetrizing" the h-variables and d-constants: for any edge $(k, j) \in E$, define $h_{k,j} = h_{j,k}$ and $\mathbf{d}_{k,j} = \mathbf{d}_{j,k}$.

As discussed in [10], we can see that for any cycle in G_i, the summation of all the h-variables corresponding to the edges on the cycle is a constant. The constant is said to be *associated* with the *cycle*. Formally, given a cycle $C = j_0, \ldots, j_k, j_0$ in G_i, the constant b is defined as $b = \sum_{r=0}^{k} d_{j_r, j_{r+1} \bmod k+1}[i]$. We can easily verify $\sum_{r=0}^{k} h_{j_r, j_{r+1} \bmod k+1} = b$ [10].

Moreover, if the p-variables at the endpoints of a path are pre-determined, then the summation of all the h-variables corresponding to the edges on the path is a constant. The constant is said to be *associated* with the *path* [10]. Formally, given

---

[c]Conventionally, k is called the *father* of j in graph theory. In order to distinguish the concept of father in graph theory and the concept of biological father, we use the term *predecessor* in this paper.

a path $\mathcal{P} = j_0, \ldots, j_k$ in G_i connecting vertices j_0 and j_k, if the variables $p_{j_0}[i]$ and $p_{j_k}[i]$ are pre-determined, the binary constant b is defined as $b = p_{j_0}[i] + p_{j_k}[i] + \sum_{r=0}^{k-1} d_{j_r,j_{r+1}}[i]$. It is easy to verify that $\sum_{r=0}^{k-1} h_{j_r,j_{r+1}} = b$ [10]. Again, notice that the constant b does not depend on the direction that path $\mathcal{P}$ is read.

2.3. *Linear Constraints on the h-Variables*

We will introduce constraint equations to "cover" all the edges in each locus graph. As mentioned above, these equations connect the p-variables and will suffice to help determine their values. The constraints can be classified into two categories with respect to the spanning tree $\mathcal{T}(G)$: constraints for cross edges and constraints for tree edges.

Cross Edge Constraints. Adding a cross edge e to the spanning tree $\mathcal{T}(G)$ yields a cycle $\mathcal{C}$ in the pedigree graph G. Suppose that the edge e exists in the i-th locus graph G_i, and consider two cases of the cycle $\mathcal{C}$ with respect to graph G_i.

Case 1: The cycle exists in G_i. We introduce a constraint along the cycle, which is called a *cycle constraint*. The set of such cycle constraints for edge e in all locus graphs is denoted by $C^{\mathrm{c}}(e)$, *i.e.*

$$C^{\mathrm{c}}(e) = \{(b, e) \mid b \text{ is associated with the cycle in } \mathcal{T}(G_i) \cup \{e\}, 1 \le i \le m\}$$

The set of cycle constraints for all cross edges is denoted by $C^{\mathrm{c}} = \biguplus_{e \in E^{\mathrm{x}}} C^{\mathrm{c}}(e)$. [d]

Case 2: Some of the edges of the cycle do not exist in G_i. This means that the cycle $\mathcal{C}$ is broken into several disjoint paths in G_i by the pre-determined vertices. Since e exists in G_i, exactly one of these paths, denoted as $\mathcal{P}$, contains e. Observe that both endpoints of $\mathcal{P}$ are pre-determined and the constraint concerning the h-variables along the path. Such a constraint will be called a *path constraint*. The set of such path constraints for e in all locus graphs G_i is denoted by $C^{\mathrm{p}}(e)$, *i.e.*

$$C^{\mathrm{p}}(e) = \left\{ (k, j, b, e) \,\middle|\, \begin{array}{l} \text{in } \mathcal{T}(G_i) \cup \{e\}, b \text{ is associated with the path containing } e \\ \text{connecting two pre-determined vertices } k \text{ and } j, 1 \le i \le m \end{array} \right\}$$

The set of path constraints for all cross edges is denoted by $C^{\mathrm{p}} = \biguplus_{e \in E^{\mathrm{x}}} C^{\mathrm{p}}(e)$.

Tree Edge Constraints. There is an implicit constraint concerning the h-variables along each path between two pre-determined vertices in the same connected component of $\mathcal{T}(G_i)$. Therefore, for each connected component of $\mathcal{T}(G_i)$, we arbitrarily pick a pre-determined vertex in the component as the *seed* vertex, and generate a constraint for the unique path in $\mathcal{T}(G_i)$ between the seed and each of the other pre-determined vertices in the component. Such a constraint will be called a *tree constraint*. Notice that if there exists any component having no pre-determined vertices, then locus i must be heterozygous across the entire pedigree and $\mathcal{T}(G_i)$ is actually a spanning tree. To conform with the notation of path constraints and for the convenience of presentation, we arbitrarily pick a tree edge denoted as e_0, and write the set of tree constraints at all loci as

$$C^{\mathrm{T}} = \left\{ (k, j, b, e_0) \,\middle|\, \begin{array}{l} \text{in a connected component of } \mathcal{T}(G_i) \text{ with seed } k, b \text{ is associated with} \\ \text{the path connecting vertices } k \text{ and a predetermined vertex } j, 1 \le i \le m \end{array} \right\}$$

Note that e_0 is the same for all the tree constraints, and will be used as an *indictor* to distinguish tree constraints from path constraints defined by cross edges.

---

[d]Here the operator $\biguplus$ denotes disjoint union and is used to save running time in our algorithm.

Again, we need symmetrize path constraints and tree constraints: given any constraint (k, j, b, e) generated for a path connecting two pre-determined vertices k and j in a locus graph, define $(k, j, b, e) = (j, k, b, e)$. We can easily see that $|C^{\mathrm{C}}| + |C^{\mathrm{P}}| + |C^{\mathrm{T}}| \le 2mn$, since $|C^{\mathrm{P}}| + |C^{\mathrm{C}}| \le m|E^{\mathrm{X}}| = m(n+1)$ and $|C^{\mathrm{C}}| \le m(|V| - 1) = m(n-1)$. Moreover, the construction of the locus graphs, the constraints C^{C}, C^{P} and C^{T} can be done in $O(mn)$ time (the detailed pseudo-code for the constructions is given in [10]).

3. A Linear-Time Algorithm for Solving Loop-Free ZRHC

As pointed out in [10], ZRHC has a solution if and only if the above linear system for the h-variables has a solution. Now we focus on how to solve the linear system to obtain solutions to the h-variables. We will check the consistency of the path/cycle constraints and tree constraints first, perform some transformation on the constraints to obtain an equivalent but smaller system of constraints, and then solve the h-variables for tree edges and cross edges with respect to the spanning tree $\mathcal{T}(G)$.

3.1. *Consistency Checking*

First, we transform the path/cycle constraints to eliminate redundancy. The tree constraints will be dealt with in a similar way.

Path/Cycle Consistency Checking and Redundancy Removal. Due to page constraint, we leave the proof of the following lemma and corollary in the full version [1]. The pseudo-code for path/cyle constraint consistency checking is given in Fig. 3.

Lemma 3.1. *Given the path constraint set $C^{\mathrm{P}}(e)$ and the cycle constraint cycle $C^{\mathrm{C}}(e)$ for a cross edge e in a tree pedigree, we can reduce them to an equivalent constraint set of size at most one, in $O(|C^{\mathrm{C}}(e) \uplus C^{\mathrm{P}}(e)|)$ time.*

As a natural extension of Lemma 3.1, we have the following corollary:

Corollary 3.1. *Given the path constraint set C^{P} and the cycle constraint cycle C^{C} for a tree pedigree, we can reduce them to an equivalent constraint set of size at most $|E^{\mathrm{X}}|$, in $O(|C^{\mathrm{C}} \uplus C^{\mathrm{P}}|)$ time.*

Tree Constraint Consistency Checking and Redundancy Removal. To clearly demonstrate the relationship among the constraints in C^{T}, we define a *constraint graph* G^* as follows. [e] For each vertex in the pedigree graph, we create a vertex in G^*. For each tree constraint $(k, j, b, e_0) \in C^{\mathrm{T}}$, we introduce an edge connecting vertices k and j in G^* with weight b. We denote by $E(G^*)$ the edge set in G^*. The weight of a path $\mathcal{P}$ (or a cycle $\mathcal{C}$) in the constraint graph is defined as the summation (in $F(2)$) over the weights of the edges along path $\mathcal{P}$ (or cycle $\mathcal{C}$), which is denoted as $W(\mathcal{C})$ (or $W(\mathcal{P})$, respectively). We choose a spanning forest $\mathcal{T}(G^*)$ on the constraint graph G^*. In each connected component $\mathcal{T}$ of $\mathcal{T}(G^*)$, the vertex with the smallest index is designated as the root of $\mathcal{T}$. (We set the roots in this way to facilitate solving the h-variables in subsection 3.2). $\mathcal{T}(G^*)$ partitions the edge set $E(G^*)$ into two subsets: *tree edges* and *cross edges*. Conventionally, a cycle containing exactly one cross edge is called a *basic cycle*. Figure 5 illustrates an example constraint graph and its basic cycles.

Procedure CONSTRAINTS_CONSISTENCY
input: $C^{\mathbb{C}}$, $C^{\mathbb{P}}$, $C^{\mathbb{T}}$, and a fixed tree edge e_0
output: compact $C^{\mathbb{C}}, C^{\mathbb{P}}$ and $C^{\mathbb{T}}$

begin

Step 1. *Consistency checking for path/cycle constraints*
  for each cross edge e
    Pick an arbitrary constraint, say (j,k,b,e), from $C^{\mathbb{C}}(e)$;
    if there exists a constraint $(j,k,b+1,e)$ $\in C^{\mathbb{C}}(e)$
      exit "Input genotypes are inconsistent.";
    $C^{\mathbb{P}} = \{(j,k,b,e)\}$;

  for each cross edge e
    Pick an arbitrary constraint, say (b,e), from $C^{\mathbb{C}}(e)$;
    if there exists a constraint $(b+1,e) \in C^{\mathbb{C}}(e)$
      exit "Input genotypes are inconsistent.";
    $C^{\mathbb{C}} = \{(b,e)\}$;
    if $C^{\mathbb{P}} \neq \emptyset$ (*i.e.* $C^{\mathbb{P}}$ contains exactly one constraint)
      Suppose the constraint $\in C^{\mathbb{P}}$ is (j,k,b',e);
      $C^{\mathbb{T}} = \{(j,k,b'+b,e_0)\}$;
      $C^{\mathbb{P}} = \emptyset$;

Step 2. *Consistency checking for tree constraints*
Construct the constraint graph G^* for $C^{\mathbb{T}}$;
$\widehat{C} = \emptyset$;

for each connected component S of G^*
Generate the spanning tree $\mathcal{T}(S)$ of S;
Suppose vertex k_0 is the vertex with the smallest index;
Let vertex k_0 be the root of $\mathcal{T}(S)$;

Traverse $\mathcal{T}(S)$ by BFS starting from k_0;
for each vertex $k \in S$
  $visited(k) = false$;
$W(k_0) = 0$; $visited(k_0) = true$;
for each tree edge (j,k) with respect to $\mathcal{T}(S)$
  Suppose the edge represents constraint (j,k,b,e_0)
  Suppose $visited(j)=true, visited(k)=false$;
  $W(k) = W(j) + b$; $visited(k) = true$;
  $\widehat{C} = \widehat{C} \uplus \{(k_0,k,W(k),e_0)\}$;

for each non-tree edge (j,k) w.r.t. $\mathcal{T}(S)$
  Suppose the edge represents constraint (j,k,b,e_0)
  if $W[j] + W[k] + b = 1$ (*i.e.* $\exists$ a 1-weight basic cycle)
    exit "Input genotypes are inconsistent.";
$C^{\mathbb{T}} = \widehat{C}$;

return $C^{\mathbb{C}}, C^{\mathbb{P}}$ and $C^{\mathbb{T}}$
end.

Algorithm LOOP_FREE_ZRHC_PHASE
input: a tree pedigree $G = (V,E)$ and genotypes $\{\mathbf{g}_j\}$
output: a general solution of $\{\mathbf{p}_j\}$
begin

Step 1. *Preprocessing*
  Choose an arbitrary vertex r as the root of a spanning tree $\mathcal{T}(G)$ for the pedigree graph G;
  Index the vertices by a BFS traversal on $\mathcal{T}(G)$;

Step 2. *Constraint generation*
  e_0 is an arbitrary tree edge w.r.t. $\mathcal{T}(G)$;
  Generate $C^{\mathbb{C}}$, $C^{\mathbb{P}}$ and $C^{\mathbb{T}}$;

Step 2'. *Redundant Constraint Elimination*
  CONSTRAINTS_CONSISTENCY($C^{\mathbb{C}}, C^{\mathbb{P}}, C^{\mathbb{T}}, e_0$);
  Define $C = C^{\mathbb{C}} \uplus C^{\mathbb{P}} \uplus C^{\mathbb{T}}$;

Step 3. *Solve the h-variables*
Construct the mapping $f : C \mapsto E'$ in Equ.(3);
Traverse G in BFS order;
$W[r] = 0$;
for each tree edge $(pred(k),k) \in E$
  if $(pred(k),k) \notin E'$
    Define $h_{pred(k),k}$ as a free variable;
    $W(k) = h_{pred(k),k} + W(pred(k))$;
  if $(pred(k),k) \in E'$
    Suppose $f^{-1}((pred(k),k))=(j,k,b,e_0)$ $(index(j)<index(k))$;
    $h_{pred(k),k} = b + W(j) + W(pred(k))$;
    $W(k) = b + W(j)$;

for each cross edge $(i,j) \in E$
  if $(i,j) \notin E'$
    Define $h_{i,j}$ as a free variable;
  if $(i,j) \in E'$
    if $f^{-1}((i,j))$ is a path constraint, say, (k_1,k_2,b,e)
      $h_{i,j}=b+W(k_1)+W(k_2)+W(i)+W(j)$;
    else (*i.e.* $f^{-1}((i,j))$ is a cycle constraint, say, (b,e))
      $h_{i,j} = b + W(i) + W(j)$;
Step 4. *Solve $\{\mathbf{p}_j\}$ by propagation*
  for each locus i
    for each component $\mathcal{T}$ in $\mathcal{T}(G_i)$
      if $\mathcal{T}$ has no pre-determined vertices
        Set the seed's p-variable as a free variable and treat it as determined;
      Traverse $\mathcal{T}$ by BFS from the seed;
      for each edge (j,k) in $\mathcal{T}$
        if $p_j[i]$ is determined but $p_k[i]$ is undetermined
          $p_k[i] = p_j[i] + h_{j,k} + d_{j,k}[i]$;

return $\{\mathbf{p}_j\}$;
end.

Fig. 3. The procedure for checking the consistency among the constraints.

Fig. 4. The algorithm for solving the linear system for Loop-Free ZRHC.

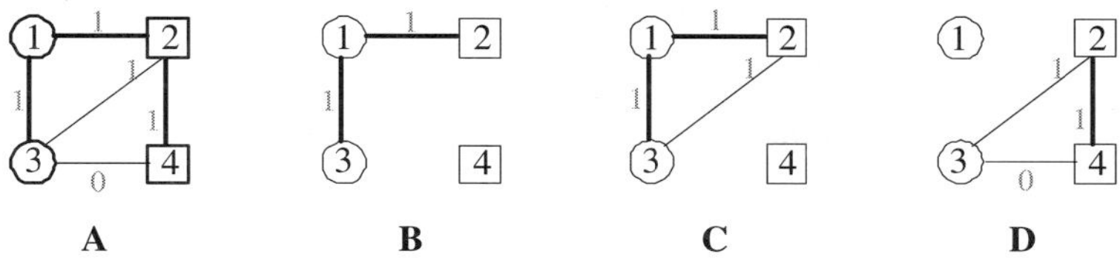

A   **B**   **C**   **D**

Fig. 5.   A constraint graph and its basic cycles. (A). Given a set of tree constraints $C^{\mathrm{T}} = \{(v_1, v_2, 1, e_0), (v_1, v_3, 1, e_0), (v_2, v_3, 1, e_0), (v_2, v_4, 1, e_0), (v_3, v_4, 0, e_0)\}$, we can construct the constraint graph G^* shown in (A), where the number on an edge in G^* is the weight of the edge, *i.e.* the b-constant in the constraint represented by the edge. For example, $(v_1, v_3, 1, e_0)$ is represented by edge (v_1, v_3) with weight 1 in the constraint graph G^*. We choose a spanning tree $T(G^*)$, where the tree edges of $T(G^*)$ are highlighted. (B). A path connecting v_2 and v_3 on the spanning tree $T(G^*)$. The weight of the path (*i.e.* the summation over weights of the edges along the path) is 0.   (C). A cycle consisting of v_1, v_2 and v_3 with weight 1. The cycle contains only one cross edge in $E(G^*)$, and is thus a basic cycle. (D). A cycle consisting of v_2, v_3 and v_4 with weight 0. The cycle contains two cross edges in $E(G^*)$, and is thus not a basic cycle.

In the rest of this subsection, we will consider some basic properties of the constraint graph, which will be used to check the consistency on tree constraints. The proofs of the following lemmas are given in the appendix.

Lemma 3.2. *For any three vertices i, j and k in a connected component of the spanning forest $T(G^*)$, the weight of the path connecting vertices j and k is equal to the total weight of the path connecting vertices i and j and the path connecting vertices i and k.*

Lemma 3.3. *In a constraint graph, the weight of every cycle can be represented as a summation over the weights of some basic cycles.*

The next corollary follows from Lemma 3.3 easily.

Corollary 3.2. *In a constraint graph, every cycle has weight 0 if and only every basic cycle has weight 0.*

Now we present our key lemma, which suggests how to use the constraint graph constructed from C^{T} to check the consistency among the constraints in C^{T}.

Lemma 3.4. *C^{T} has a solution if and only if every cycle in the corresponding constraint graph has weight 0.*

Corollary 3.2 and Lemma 3.4 imply the following theorem.

Theorem 3.1. *C^{T} has a solution if and only if every basic cycle in the corresponding constraint graph has weight 0.*

Therefore, we can check the consistency of C^{T} by detecting the existence of 1-weight basic cycles. Notice that every basic cycle contains exactly one cross edge in $E(G^*)$. We can calculate the weights of basic cycles by first calculating the weight of the path connecting every pair of vertices in each connected component of the spanning forest $T(G^*)$, and then checking if the basic cycle induced by each cross edge has weight 0. By Lemma 3.2, in every connected component T of the forest $T(G^*)$, the weight of the path connecting each pair of vertices can be obtained from the weights of the paths connecting the root of T and each of involved vertices. In the following, we denote by $W(k)$ the weight of the path between the root and a vertex k in a connected component of $\mathcal{T}(G^*)$.

The detailed pseudo-code for checking the consistency of $C^\mathtt{T}$ is given in Figure 3. Besides consistency checking, the procedure also generates an equivalent constraint set $\widehat{C}$ for $C^\mathtt{T}$ based on $\mathcal{T}(G^*)$, which is defined as follows,

$$\widehat{C} = \left\{ (k_0, k, W(k), e_0) \,\middle|\, \begin{array}{l} k \in V, \ k_0 \text{ is the root of the connected component} \\ \text{containing } k \text{ in } \mathcal{T}(G^*), \text{ and } k \neq k_0 \end{array} \right\} \qquad (2)$$

Clearly, $|\widehat{C}| \leq |V| - 1 = n - 1$. This discussion is summarized formally in the following lemma. [f]

Lemma 3.5. *Given the tree constraint set $C^\mathtt{T}$, we can reduce it to an equivalent constraint set of size at most $(n-1)$ in $O(|C^\mathtt{T})|$ time.*

The following corollary holds from Corollary 3.1, Lemma 3.5, and the fact $|C^\mathtt{C}| + |C^\mathbb{P}| + |C^\mathtt{T}| \leq 2mn$.

Corollary 3.3. *Given the linear constraint set $C = C^\mathtt{C} \uplus C^\mathbb{P} \uplus C^\mathtt{T}$ for a tree pedigree, we can reduce it to an equivalent constraint set of size at most $|E|$ in $O(mn)$ time, where E is the set of the edges in the pedigree graph and $|E| \leq 2n$.*

3.2. *Solving the h-Variables Efficiently*

Now we return to the pedigree graph G and show how to solve the h-variables for tree edges and for cross edges separately. Our idea is as follows. We first construct a one-to-one mapping from the constraints to the edges in the pedigree graph, and then assign values to the h-variables consistent with the constraints by a BFS traversal on the pedigree graph.

Constructing a Mapping from Constraints to Edges. We want to construct a mapping f from the (reduced) constraints $C = C^\mathtt{C} \uplus C^\mathbb{P} \uplus C^\mathtt{T}$ to a subset $E' \subseteq E$ of the edges satisfying the following two conditions:

- *Condition 1:* For every tree/path constraint (or every cycle constraint), there exists exactly one edge of E' on the path (or the cycle, respectively) corresponding to the constraint.
- *Condition 2:* f is one-to-one.

We define the mapping $f : C \mapsto S'$ as follows, where $C = C^\mathtt{T} \uplus C^\mathbb{P} \uplus C^\mathtt{C}$ and the subset S' is implied by the definition.

$$f(c) = \begin{cases} (pred(k), k) & \text{if } c = (j, k, b, e_0) \in C^\mathtt{T} \text{ and } index(j) < index(k) \ (i.e.\ c \text{ is a tree constraint}) \\ (i, j) & \text{if } c = (k_1, k_2, b, e) \in C^\mathbb{P}, \text{ where } e = (i, j) \ (i.e.\ c \text{ is a path constraint}) \\ (i, j) & \text{otherwise } c = (b, e) \in C^\mathtt{C}, \text{ where } e = (i, j) \ (i.e.\ c \text{ is a cycle constraint}) \end{cases}$$
$$(3)$$

Clearly, the mapping for every path (or cycle) constraint satisfies condition 1. Consider a tree constraint $(j, k, b, e_0) \in C^\mathtt{T}$ and suppose that $index(j) < index(k)$. Because we construct the indexing from top to bottom in $\mathcal{T}(G)$, vertex j must be at the same or a higher level than vertex k. Hence, $(pred(k), k)$ must be on the path connecting vertices j and k in $\mathcal{T}(G)$. As a result, condition 1 holds for the mapping f. The proof of the mapping f is one-to-one is omitted here and can be found in the full version [1].

---

[f]A similar lemma is given in [10] too.

Once the mapping f is defined, we can solve the constraints as follows. We set each h-variable corresponding to an edge in $E - E'$ as a free variable, and represent the h-variable corresponding to an edge in E' as a linear combination of the free variables based on the constraint mapped to the edge so that the constraint is satisfied.

Assigning Values to the h-Variables We need assign values to the h-variables so that all the path/cycle/tree constraints are satisfied. Given a vertex j, we denote by $W[j]$ the sum of the h-variables along the path between the root of the spanning tree $\mathcal{T}(G)$ and vertex j. We will treat h-variables for tree edges and cross edges separately, utilizing the mapping f.

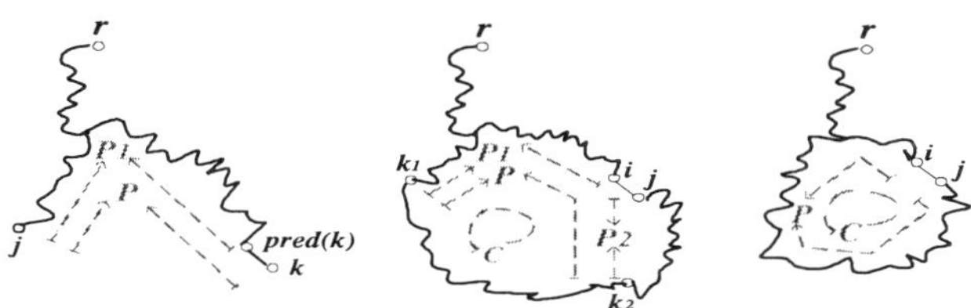

Fig. 6. The spanning tree $\mathcal{T}(G)$ with root r. The tree edges are highlighted. (A) The path $\mathcal{P}$ connecting vertices j and k on $\mathcal{T}(G)$ for a given tree constraint (j, k, b, e_0), and the path $\mathcal{P}_1$ connecting vertices j and $pred(k)$. (B) The cycle $\mathcal{C}$ induced by adding $e = (i, j)$ to the spanning tree $\mathcal{T}(G)$ for given a path constraint (k_1, k_2, b, e), and the paths $\mathcal{P}$, $\mathcal{P}_1$, and $\mathcal{P}_2$ connecting vertices k_1 and k_2, vertices k_1 and i, and vertices j and k_2, respectively. Notice that $\mathcal{P}_1$ and $\mathcal{P}_2$ do not contain e while $\mathcal{P}$ does. (C) The cycle $\mathcal{C}$ induced by adding $e = (i, j)$ to the spanning tree $\mathcal{T}(G)$ for a given cycle constraint (b, e) and the path $\mathcal{P}$ connecting the vertices i and j. Note that $\mathcal{P}$ does not contain e.

- **Handling h-variables for tree edges.** We solve the h-variables by a BFS traversal on $\mathcal{T}(G)$ starting from the root r. First, we set $W[r] = 0$. If we encounter a tree edge $(pred(k), k)$ in $(E - E')$, we set $h_{pred(k),k}$ as a free variable, and set $W(k) = h_{pred(k),k} + W(pred(k))$. Otherwise, we encounter a tree edge $(pred(k), k)$ in E' (assuming $f^{-1}((pred(k), k)) = (j, k, b, e_0)$ and $index(j) < index(k)$), and we set $h_{pred(k),k}$ as:

$$h_{pred(k),k} = b + W(j) + W(pred(k)) \tag{4}$$

  Note that since $index(j) < index(k)$ and $index(pred(k)) < index(k)$, $W(j)$ and $W(pred(k))$ are known at this time. We also set $W(k) = b + W(j)$. An illustration of the vertices j, $pred(k)$ and k is given in Figure 6(A). After the BFS traversal, we determine $W[j]$ for every vertex j. We will use $W[j]$ to solve the h-variables for cross edges.

- **Handling h-variables for cross edges:** If we encounter a cross edge $(i, j) \in (E - E')$, we set $h_{i,j}$ as a free variable. Otherwise, we encounter a cross edge $(i, j) \in E'$, and have $|C^c(e) \uplus C^{\mathbb{P}}(e)| = 1$ based on Lemma 3.3 and Equation (3). We assign $h_{i,j}$ as follows.

  - *Case 1:* $C^c(e) \uplus C^{\mathbb{P}}(e)$ contains a path constraint (k_1, k_2, b, e). We set

$$h_{i,j} = b + W(k_1) + W(k_2) + W(i) + W(j) \tag{5}$$

  An illustration of the vertices k_1, k_2, i and j is demonstrated in Figure 6(B).

– *Case 2:* $C^c(e) \uplus C^{\mathbb{P}}(e)$ contains a cycle constraint (b, e). We set

$$h_{i,j} = b + W(i) + W(j) \tag{6}$$

An illustration of the vertices i and j is given in Figure 6(C).

Figure 4 displays a pseudo-code of the above procedure. The algorithm is called **Loop_Free_ZRHC_Phase**, which generates a general solution. Note that if we assign a particular binary value to each free variable in the algorithm, the algorithm obtains a particular solution. The correctness of the algorithm is given in the following lemma.

Lemma 3.6. *The algorithm* **Loop_Free_ZRHC_Phase** *correctly solves the ZRHC problem on a tree pedigree.*

Theorem 3.2. *The algorithm* **Loop_Free_ZRHC_Phase** *produces a particular solution to the ZRHC problem on a tree pedigree in* $O(mn)$ *time, and a general solution in* $O(mn^2)$ *time.*

4. Concluding Remarks

It still remains open if one could extend the algorithm **Loop_Free_ZRHC_Phase** to solve the ZRHC problem on a general pedigree in linear time.

Acknowledgement

We would like to thank Jing Xiao, Marek Chrobak and Qi Fu for many useful discussions. The research is supported in part by NSF grant CCR-0309902, NIH grant LM008991-01, NSFC grant 60528001, and the Changjiang Visiting Professorship at Tsinghua University.

References

[1] The full version of this paper is available at `http://www.cs.ucr.edu/~lliu/paper/tree_zrhc_full.ps`

[2] G. R. Abecasis, S. S. Cherny, W. O. Cookson and L. R. Cardon, Merlin–rapid analysis of dense genetic maps using sparse gene flow trees. *Nat. Genet.*, 30(1):97-101, 2002.

[3] J. R. O'Connell. Zero-recombinant haplotyping: applications to fine mapping using SNPs. *Genet. Epidemiol.*, 19 Suppl 1:S64–70, 2000.

[4] M. Y. Chan, W. Chan, F. Chin, S. Fung, and M. Kao Linear-Time Haplotype Inference on Pedigrees without Recombinations. *Proc. of the 6th Annual Workshop on Algorithms in Bioinformatics (WABI'06)*, 56-67 ,2006.

[5] The international HapMap Consortium. International HapMap Project. *Nature*, 426:789-796, 2003.

[6] J. Li and T. Jiang. Efficient rule-Based haplotyping algorithm for pedigree data. *Proc. of 7th Annual Conference on Research in Computational Molecular Biology (RECOMB'03)*, 197-206, 2003.

[7] J. Li and T. Jiang. An exact solution for finding minimum recombinant haplotype configurations on pedigrees with missing data by integer linear programming. *Proc. of RECOMB'04*, 20-29, 2004.

[8] J. Li and T. Jiang. Computing the minimum recombinant haplotype configuration from incomplete genotype data on a pedigree by Integer Linear Programming. *J. of Computational Biology* 12(6), 719-739, 2005.

[9] D. Qian and L. Beckmann. Minimum-recombinant haplotyping in pedigrees. *Am. J. of Hum. Genet.*, 70(6):1434–1445, 2002.

[10] J. Xiao, L. Liu, L. Xia, T. Jiang. Fast Elimination of Redundant Linear Equations and Reconstruction of Recombination-Free Mendelian Inheritance on a Pedigree. *Proc. of 18th Annual ACM-SIAM Symoposium on Discrete Algorithms (SODA'07)*, 655-664, 2007.

COMPUTED PROTONATION PROPERTIES: UNIQUE CAPABILITIES FOR PROTEIN FUNCTIONAL SITE PREDICTION

LEONEL F. MURGA[1,2]   YING WEI[1]
leonel@brandeis.edu   wei.y@neu.edu

MARY JO ONDRECHEN[1]
mjo@neu.edu

[1]*Department of Chemistry & Chemical Biology and Institute for Complex Scientific Software, Northeastern University, Boston, MA 02115 USA*
[2] *Present Address: Rosenstiel Basic Medical Sciences Research Center, Brandeis University, Waltham, MA 02454 USA*

Prediction of protein functional sites from 3D structure is an important problem, particularly as structural genomics projects produce hundreds of structures of unknown function, including novel folds and the structures of orphan sequences. The present paper shows how computed protonation properties provide unique and powerful capabilities for the prediction of catalytic sites from the 3D structure alone. These protonation properties of the ionizable residues in a protein may be computed from the 3D structure using the calculated electrical potential function. In particular, the shapes of the theoretical microscopic titration curves (THEMATICS) enable selection of the residues involved in catalysis or small molecule recognition with good sensitivity and precision. Results are shown for 169 annotated enzymes in the Catalytic Site Atlas (CSA). Performance, as measured by residue recall and precision, is clearly better than that of other 3D-structure-based methods. When compared with methods based on sequence alignments and structural comparisons, THEMATICS performance is competitive for well-characterized enzymes. However THEMATICS performance does not degrade in the absence of similarity, as do the alignment-based methods, even if there are few or no sequence homologues or few or no proteins of similar structure. It is further shown that the protonation properties perform well on open, unbound structures, even if there is substantial conformational change upon ligand binding.

Keywords: function prediction; site prediction; structural genomics; THEMATICS; electrostatics

1.   Introduction

Structural genomics efforts are revealing the 3D structures of hundreds of proteins of unknown function, including the structures of orphan sequences and structures with novel features or novel folds. A recent search of the Protein Data Bank (PDB) [4] returned 3108 structures listed as "hypothetical" or "unknown function" or "putative." This number is constantly increasing at a rapid pace as techniques for high-throughput expression and crystallization are further developed and refined. However, knowledge of the three-dimensional structure does not necessarily imply knowledge of function. In fact, the inference of functional information from the 3D structure has proved to be far more difficult than anticipated. Computational methodologies for the prediction of functional information about these proteins are therefore important and timely.

An additional challenge posed by structural genomics proteins is that nearly all of them are unbound structures, since their natural substrates and other ligands are generally not present or even known. Ligand binding is usually accompanied by some structural change. While these binding-induced structural changes are often small and involve only rotation of the side chains of adjacent residues, significant change in backbone conformation does occur for some proteins. Thus it is desirable to develop predictive structure-to-function methods that are successful for these unbound, apo structures.

Many methods have been reported to date for the prediction of functional sites in proteins. Some of these methods utilize only the 3D structure of the query protein as input [1, 3, 12, 13, 16], and thus are applicable to proteins with few or no sequence homologues, while other methods require both the 3D structure and a sequence alignment [6, 15, 17, 18, 23]. These methods report high success rates in functional site prediction, although often the precision is low, with only a small fraction of the selected residues corresponding to annotated functional residues. Achieving high recall with good precision is a challenge but yields more useful results.

The reliability of computed protonation properties for the identification of interaction sites in protein 3D structures was first reported in 2001 [16]. The method was named THEMATICS (Theoretical Microscopic Titration Curves) and has since been automated using statistical metrics [12, 21]. In the present paper it is shown how computed protonation properties can predict the interaction sites in open, unbound structures for systems undergoing a large conformational change upon ligand binding. It is also shown that these properties alone return low false positive rates, in addition to high site success rates and good residue recall, for the enzymes in an annotated dataset. Examples of the precise, highly localized, predictions are given, including cases with large apo-holo conformational change.

2.   Method

Protein structure coordinate files are downloaded from the PDB. The 3D structure files are pre-processed as described by Wei [21]. From the atomic coordinates, the electrical potential function of the protein is calculated with a Finite Difference Poisson–Boltzmann (FDPB) procedure. The FDPB component of the University of Houston Brownian Dynamics (UHBD) program [14] is used for this purpose. The theoretical titration curves are calculated for each ionizable species in the protein (each Arg, Asp, Cys, Glu, His, Lys, and Tyr, plus the N- and C- termini) using a hybrid procedure [10]. Each of these chemical groups in a protein structure can gain or lose an H^+ ion, a proton. For each of these ionizable groups, the hybrid procedure computes the fraction of all the protein molecules in a large ensemble that have this group occupied by a proton at a given pH. These data are expressed as the proton occupation O as a function of the pH. These O(pH) curves constitute theoretical analogues of an experimental titration curve. We have argued that catalytic residues have special properties in their proton transfer chemistry that can be observed in the shapes of the computed O(pH) curves [16].

An extension of Ko's analysis [12] is used to select the residues that are most likely to be involved in catalysis and/or recognition. We define the first derivative function f of the O(pH) curve as:

$$f = -\, dO/d(pH) \tag{1}$$

These f functions are essentially proton binding capacities [8, 9, 22], which measure the change in concentration of a bound proton per unit change in its chemical potential. Note that these f functions are automatically normalized so that the area under the f curve is unity. This is because O always runs from 1 to 0, so that $-\Delta O$ is always unity over the full range of pH values; this is true for both perturbed and normal ionizable residues.

The f functions may be treated as distributions and characterized by their moments [8, 12]. Hence we define the nth central moment μ_n as:

$$\mu_n = \int (pH - M_1)^n \cdot f \cdot d(pH) \tag{2}$$

where M_1 is the first raw moment, defined by the expression for the nth raw moment as:

$$M_n = \int (pH)^n \cdot f \cdot d(pH) \tag{3}$$

Integrals in (2) and (3) are over all space ($-\infty$ to $+\infty$). Equations 1-3 for each ionizable residue are evaluated numerically from the computed theoretical titration curve of each residue.

If there were only one ionizable residue in the protein, that residue would obey the Henderson-Hasselbalch (H-H) equation. For such a residue that obeys the H-H equation, the first raw moment is the pK_a and all the odd-numbered central moments are zero. The second and fourth central moments have the values 0.620 and 1.62, respectively, for an H-H acid or base. However, interactions between ionizable residues in a protein will lead to asymmetry and broadening of the f functions and thus the odd-numbered central moments will be non-zero and the even-numbered moments will be larger. The underlying premise of THEMATICS is that these interactions are strongest for the active site residues and therefore the active site residues are identifiable as those with the most deviant curve shapes and particularly the largest third and fourth moments [12].

We define the Z score for the nth central moment as:

$$Z_n = (|\mu_n| - <|\mu_n|>)/\sigma_n \tag{4}$$

Here $<|\mu_n|>$ is the mean of the absolute value of the nth central moment, averaged over all of the ionizable residues in the protein. Generally only the odd-numbered moments can be negative, so the absolute value sign really is needed only for the odd moments. σ_n is the standard deviation of the (absolute) values of the nth central moment for the set of all ionizable residues in the protein. The Z score represents the deviation from the mean value in units of the standard deviation. The f functions are peaked functions for ordinary residues, but the active site residues deviate the most from the sharply peaked H-H form. The central moments are natural metrics to characterize the width and the shape of these peaked functions and their Z scores provide a way to identify the most deviant curves.

The mean and the standard deviation in Equation (4) are obtained using a specified cut-off fraction of the complete set of ionizable residues in the protein of interest. In particular, the set of ionizable residues in the protein is rank-ordered according to the nth central moment, then the only the lowest, specified fraction is used to obtain the mean and the standard deviation. Thus extremely large values for the central moments, which can have large impact on the mean and standard deviation, are excluded for this part of the calculation. For instance, if the cut-off fraction is 0.98, all of the central moment values that are above the 98th percentile are excluded when the mean and standard deviation are computed. A cut-off fraction of 1.0 is therefore equivalent to Ko's analysis [12]. Z scores are computed for all of the ionizable residues, but the mean and the standard deviation are computed using the smaller, cut-off population.

The criterion $Z_3>1$ or $Z_4>1$ is used to select the positive residues, since most active site residues have either a large third central moment or a large fourth central moment.

Once the THEMATICS positive residues are identified, these residues are grouped into clusters based on spatial proximity. The distance between two positive residues is defined as the distance between their charge centers. A residue is placed into a cluster if it is within 9Å of any other positive residue in the cluster. A one-member cluster is called an isolated positive and it is not considered predictive. Clusters containing two or more positive residues constitute predictions and are termed THEMATICS predictive clusters.

3.   Results

3.1   *Performance in Site Prediction*

The method was tested on 169 annotated enzymes, essentially the entire CatRes database, with updated annotations from the Catalytic Site Atlas (CSA) database [2, 19]. Performance is measured by the recall (fraction of residues annotated in the database as catalytically important that are identified by the method), precision (fraction of identified residues that are annotated in the database as catalytically important), and false positive (FP) rate. We note that precision rates should be considered as lower bounds, because the database annotations are incomplete. In other words, not all of the important residues are annotated as such in the database. Recall, precision, FP rate, and Matthews Correlation Coefficient (MCC) for 169 CatRes enzymes, computed using CatRes/CSA annotations only, are shown in Table 1 for different values for the cut-off fraction. Recall rates for residues annotated as catalytically important range from 41% to 63%. Nominal precision rates range from 19% to 14%; this represents the fraction of predicted residues that are annotated in the database as catalytically important. FP rates are low, ranging from 2.0% to 4.7%, depending on the cut-off. Note that the lower cut-off values result in more residues selected, so as the cut-off value is reduced the residue recall rate rises but at some expense in precision.

While Table 1 presents performance in the prediction of catalytic residues, the success rate for the prediction of catalytic sites is higher. Table 2 shows the percentages of the 169 test proteins for which correct or partially correct predictions are made by THEMATICS. Success in site prediction is defined according to designations used in previous work [11]. A site prediction is considered *correct* if it includes half or more of the annotated catalytic residues. A prediction is considered *partially correct* if it contains at least one, but less than half, of the annotated catalytic residues. The total success rate for the prediction of sites is the sum of the correct plus partially correct predictions. In columns 2 through 4 of Table 2, the correct and partially correct predictions are determined with the CatRes/CSA annotations only. In column 5, an expanded set of annotations, including information from the SITE fields of the PDB files and from protein-specific literature references, is used. For cases where a bound structure is available, the residues in direct contact with the bound ligand(s) were added to the list. The CatRes/CSA annotations thus constitute a subset of this expanded list. All methods, including THEMATICS, perform better against this expanded list.

Table 1. Average recall, precision, false positive rate, and MCC for THEMATICS predictions of catalytic residues as functions of the cut-off fraction for the test set of 169 enzymes. Here only the CSA annotations are used as the reference set.

Cut-off	Recall	Precision	FP rate	MCC
1.00	41.1%	19.4%	2.07%	0.255
0.99	50.4%	17.9%	2.76%	0.272
0.98	54.2%	16.4%	3.24%	0.270
0.97	58.0%	15.5%	3.74%	0.270
0.96	61.0%	14.6%	4.23%	0.268
0.95	62.8%	13.6%	4.72%	0.262

Table 2. THEMATICS success rates for site prediction for the 169 enzymes in the test set. Success rate is expressed as correct, partially correct, and total, using a cut-off of 1.00, 0.99 and 0.98. In columns 2-4, only the CSA annotations are used as the reference set. In column 5, the total success rate is obtained using an expanded set of annotations.

Cut-off	Correct Site Rate versus CSA	Partially Correct Site Rate versus CSA	Total Success Rate versus CSA	Total Success Rate versus Expanded Set
1.00	48.5%	29.0%	77.5%	89.9%
0.99	59.8%	26.0%	85.8%	92.9%
0.98	66.9%	21.3%	88.2%	94.1%

For cut-off values of 1.0, 0.99, and 0.98, THEMATICS makes correct site predictions for 49%, 60%, and 64%, respectively, of the proteins in the CatRes test set, according the CSA annotations only. Total success rates, the sum of the correct and partially correct rates for site prediction, are 90%, 93%, and 94%, for the same three respective cut-off values, according to the expanded annotation set.

These performance data shown in Tables 1 and 2 compare very favorably with other methods that are based solely on the 3D structure of the query protein. Structural Analysis of Residue Interaction Graphs (SARIG) [1] is a graph theoretic approach that calculates residue contacts and identifies the residues that have the highest closeness scores to all other residues. SARIG successfully predicts 46.5% of the annotated catalytic residues for the enzymes in the CatRes database. The reported precision, however, is low; only 9.4% of the predicted residues are known catalytic residues [1]. Thus compared to THEMATICS with a cut-off of 0.99, the residue recall rates are similar but the THEMATICS precision rate is about two-fold better.

Another approach to the prediction of sites from the structure alone involves docking of small solvent molecules onto the protein surface and searching for clusters of energy minima for these molecules [7, 20]. Q-SiteFinder is a simple, fast version of this method developed by Laurie [13] and uses only a methyl group as a probe. For 90% of proteins in the test set, Q-SiteFinder returns a correct site prediction within its top three predicted sites. While precision was not reported, selectivity clearly is low. We estimate that the precision rate for residue prediction by Q-SiteFinder is only about 5%, corresponding to an average of about 60 predicted residues per protein; these estimates are based on a combination of the top three sites as the prediction, which was the basis for the reported success rates [13]. Thus THEMATICS gives comparable success rates in site prediction but with substantially better precision and lower false positive rates.

THEMATICS also performs quite well compared with methods that require sequence alignments and structural comparisons. While there are variations in the annotated sets, one can get some idea of the relative performance. One method based on sequence and structural alignments reports a catalytic residue recall rate of 47% with an FP rate of 5% [17]; THEMATICS with a cut-off of 0.99 has a similar recall rate (53%) but the FP rate is lower by almost one half (2.8%). Another study using Support Vector Machines (SVM) to predict catalytic residues from sequence conservation and structural properties reports an MCC of 0.23 [18], slightly less than the MCC for THEMATICS (0.27). Another very recent paper [23], also using SVM with features obtained from sequence alignments and structural properties, reports 57.0% catalytic residue recall with 18.5% precision, slightly better than THEMATICS; however these data were obtained on a test set of enzymes that possess sequence homologues and structural family members. As the authors point out, there will be a cost in the recall and in the precision when applied to novel folds or remote sequences. THEMATICS performance is roughly the same as that of the SVM-based sequence/structure methods, but without any sequence or structural alignments needed.

3.2 *Predicting Holo Binding Sites from Apo Structures*

For proteins that undergo a small conformational change upon ligand binding, it has been shown already that THEMATICS performs equally well for apo (unbound) structures as it does for holo (bound) structures [21]. In this section we examine pairs of structures that exhibit significant change in backbone conformation going from the apo to the holo form. For such a pair of structures, the Root Mean Square Deviation (RMSD) of the alpha carbon framework may be calculated using the expression:

$$RMSD = \sqrt{\frac{\sum_{i}^{N} d_i^2}{N}} \tag{5}$$

Where d_i represents the distance between the alpha carbon atoms of equivalent residues in the apo and holo forms and N is the number of residues in both structures.

The RMSD itself is not a good measure of the relative change in conformation when one compares pairs of proteins with a wide range of sizes because it depends on the number of residues. To obtain a better sense of the degree of conformation change within the set of proteins studied, we also include the RMSD100 [5] value defined as:

$$RMSD100 = \frac{RMSD}{1 + \ln\left(\sqrt{\frac{N}{100}}\right)} \tag{6}$$

Where RMSD is given in equation (6). RMSD100 is a size-normalized RMSD that allows a better comparison across sets with large variations in size.

To assess how these changes affect the region specifically around the binding site, the RMSD and RMSD100 values for the alpha carbon atoms of "core residues" are reported. These are defined as the set of residues with any atom located within 8 Å of any bound ligand in the holo form. Comparison of the RMSD values for the whole protein and for core residues provides a quantitative description of the effect of the conformational change on the binding site. Ten illustrative examples are given in Table 3.

These are examples from a set of 24 proteins for which apo and holo structures are available for the same species and for which the apo-holo RMSD is 1.5A or greater. RMSD values range from 1.5 to 14.7 in the full set and from 3.7 to 14.7 in the Table 3 examples. The first column gives the species and protein name. The number of residues N in the protein is given in the second column, with the number of residues in the active site core, as defined above, in parentheses. The next two columns give the RMSD and RMSD100 values for the apo-holo pair, with the corresponding values for the core residues shown in parentheses.

THEMATICS predictions for the apo and holo structures for the Table 3 examples are given in Table 4. In Table 4, the PDB code and the THEMATICS predictions are given for the apo form first for each protein, then for the holo form. Residues that belong to the same cluster are shown together in square brackets. Clusters of two or more residues constitute predictions. Clusters with only one member are reported but are not considered predictive. Known catalytic residues and residues in direct contact with the bound ligand of the holo form are shown in **boldface**.

Table 4 demonstrates how THEMATICS is able to predict binding sites in the unbound apo structures. In some cases, for example yeast guanylate kinase, *E. coli* dipeptide binding protein, and human serum transferrin, the binding residues are divided into different, spatially separated clusters in the apo structure. Upon ligand binding, these clusters come together to form a single cluster in the holo form; these predicted residues surround the bound ligand in the holo structure.

Table 3. Ten proteins with significant conformational change upon ligand binding. Number of residues N in the protein and the apo-holo RMSD and RMSD100 are given. Corresponding values for the active site core residues are shown in parentheses.

Species and Protein Name	N (N Core)	RMSD (Core)	RMSD100 (Core)
Yeast Guanylate Kinase	186 (45)	4.4 (3.4)	5.5 (9.2)
E. coli Dipeptide Binding Protein	507 (51)	12.3 (10.3)	6.8 (15.6)
Human Serum Transferrin	328 (33)	14.7 (6.7)	9.3 (15.1)
Human Glucokinase	424 (58)	10.9 (9.5)	6.3 (13.0)
Human Lactoferrin	691 (73)	8.2 (5.8)	4.2 (6.9)
E. coli L-Leucine Binding Protein	345 (51)	14.4 (9.3)	8.9 (14.1)
E. coli D-Ribose Binding Protein	271 (47)	8.7 (5.6)	5.6 (9.0)
Emericella nidulans 3-Dehydroquinate Synthase	762 (189)	3.7 (2.3)	1.8 (1.7)
E. coli Ribokinase	610 (115)	9.8 (6.4)	5.1 (5.9)
Limulus polyphemus Arginine Kinase	344 (71)	3.8 (3.1)	2.4 (3.7)

Figure 1 shows the active site region of the aligned apo and holo structures of human serum transferrin with the THEMATICS predictions for both structures. Transferrin is the generic label for a group of proteins found in a wide range of organisms. Their main function is the sequestration of iron either for storage or for transport to the cell interior. The backbones are shown as ribbons and the THEMATICS predicted residues are shown explicitly in stick form. The apo form is in light gray, the holo form dark. Upon binding of iron, there is a large displacement of the binding residue H249. The two clusters predicted by THEMATICS for the unbound apo structure, [E83, **H249**] and [**Y95, Y188, K206**] come together to form a single cluster, [Y85, **Y95, Y188, K206, H249**, D292, **K296**], in the holo structure. Note that the predicted residues surround the bound iron in the holo form, and that THEMATICS is able to identify them even in the apo form.

In the full set of 24 pairs, THEMATICS predicts correct sites for 23 of them (96%). For one case, threonine synthase from *Thermus thermophyllus*, the correct site is predicted for the holo form but not for the apo form. Somewhat surprisingly, this particular case does not have one of the larger conformational changes (RMSD=2.6;

RMSD100=1.6). Three annotated catalytic residues, [**K61**, **K116**, **R160**], fall above the cut-off for the holo form but fall short of the cut-off for the apo form.

Table 4. THEMATICS predictions for apo and holo structures of the 10 proteins listed in Table 3. The PDB codes for the apo and holo forms are given in the second column, with the apo form given first. Predictions are given in clusters, with members of the same cluster shown together in square brackets. The known binding residues are shown in **boldface**. Clusters of two or more residues constitute predictions.

Species and Protein Name	PDB ID Apo Holo	THEMATICS Predicted Clusters
Yeast Guanylate Kinase	1EX6	[Y25, Y175, K179] [**Y50**, **Y78**] [**E69**, **D100**, **D98**, E153, H162, Y156] [C95] [R135] [D170]
	1EX7	[**K14**, **Y50**, **D98**, E153, **R38**, **R41**, **Y78**, **D100**, H162, **R146**] [Y175, K179]
E. coli Dipeptide Binding Protein	1DPE	[D26, E38, D153, E482, D494, **K498**, H499, H500] [D89] [E143, D149, **D408**, D411, D413] [Y269] [**Y357**, **Y431**]
	1DPP	[Y25, D26, **Y114**, D153, Y239, **R355**, **Y357**, **D408**, H499] [D149, D413]
Human Serum Transferrin	1BP5	[E83 , **H249**] [**Y95** , **Y188** , **K206**]
	1A8E	[Y85 , **Y95** , **Y188** , **K206** , **H249** , D292 , **K296**]
Human Glucokinase	1V4T	[E40A] [**C213A** , **C220A** , **C233A** , **C252A**] [**C230A** , **C382A**] [D274A] [E339A]
	1V4S	[**C213A** , **C220A** , **C233A** , **C252A**] [**C230A** , **C382A**]
Human Lactoferrin	1CB6	[**Y92**, **Y192**, R210] [Y93, R249] [E413 , **Y435** , **Y528** , K546, **H597**]
	1LFG	[Y82 , **Y92** , **Y192** , R210 , **H253**] [Y398] [**Y435** , **Y528** , **H597**]
E. coli L-Leucine Binding Protein	1USG	[E22 , D51 , H76 , Y89] [Y72] [E90 , H113] [D121 , **Y150** , Y198 , **E226**] [H145 , D146 , E152 , E205] [K250 , Y252 , D330] [D321 , D325]
	1USK	[E22, D51, H76] [Y72] [E90, H113] [D121, H145, D146, **Y150**, E152, E205, **E226**] [D321, D325]
E. coli D-Ribose Binding Protein	1URP	[E26] [**D89**, H100, **D215**, E140, D191, E192]
	2DRI	[**D89**, **D215**, E140, D191, E192]
Emericella nidulans 3-Dehydroquinate Synthase	1NUA	[Y25, C34, K172]* [**K84A**, K161A, **E194A**, **K197A**, **K250A**, **H271A**, E278A, **H287A**, **R130B**]* [K89A, K89B] [Y134A]* [D176A] [E181B]
	1NVF	[K89A] [**R130A**, K152B, K250B, **H271B**, **H275B**, E278B, **H287B**, K356B] [K152A, K250A, K356A, **R130B**, **H271A**, E278 , **H287A**]
E. coli Ribokinase	1RKA	[**D16A**, H17A, **D67A**, H113A, **D16B**, H17B, **D67B**, H113B] [**E143A** , **D255A**]*
	1RKS	[**D16A**, H17A, **E143A**, **D255A**, **D16B**, H17B, **E143B**, **D255B**] [E188A]* [E190A]*
Limulus polyphemus Arginine Kinase	1M80	[**Y68**, **R126**, **C127**, Y134, Y145, K151, R208, E224, **E225**, D226, H227, **R229**, **C271**, **R280**, R330] [D71]
	1BG0	[Y89, **R126**, **C127**, R208, E224, **E225**, D226, **R229**, **R28**, **R309**, **E314**, H315, R330, E335] [H185]

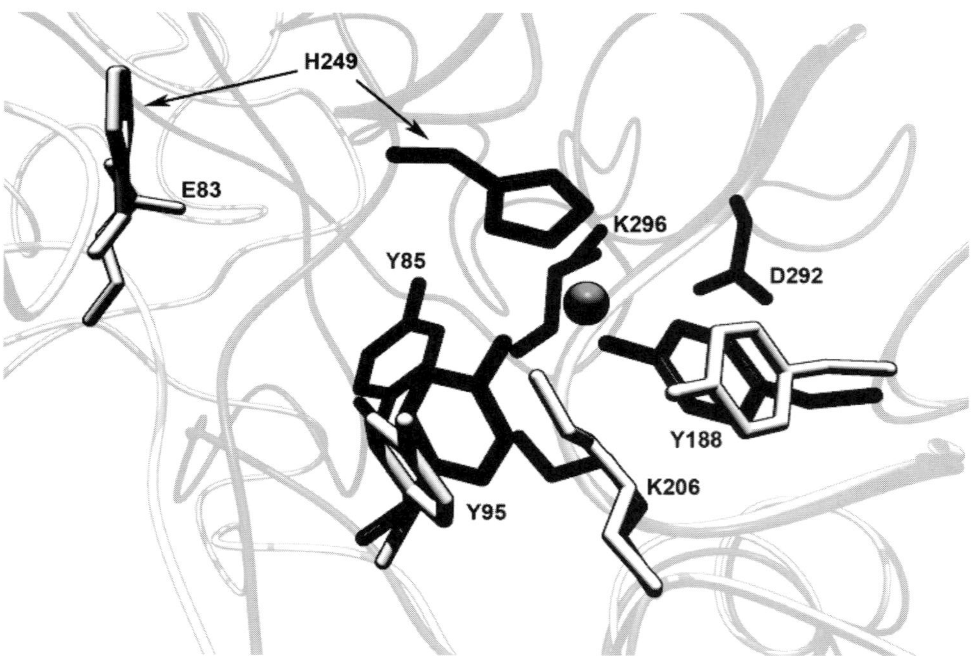

Figure 1. Detail of the active site of human serum transferrin showing the rearrangement that occurs upon binding of iron (unlabeled sphere). The apo form (PDB 1BP5) is shown in light gray; the holo form (PDB 1A8E) is in dark gray. Side chains of the THEMATICS predictions for the unbound apo form (in white), [E83, **H249**] and [**Y95**, **Y188**, **K206**] and for the bound holo form (in black), [Y85, **Y95**, **Y188**, **K206**, **H249**, D292, **K296**] are shown in stick form. Note that the two clusters in the apo structure become a single cluster in the holo form, reflecting the large displacement of H249. Figure prepared with Yasara.

4.   Discussion and Conclusions

THEMATICS requires only the 3D structure of the query protein as input and therefore its performance is not affected by the degree of sequence or structural similarity to other proteins. THEMATICS performs significantly better than other 3D-structure-based methods. THEMATICS, using only the protonation properties computed from the structure, also performs quite well against the latest SVM-based methods that do utilize sequence alignments and structural similarity; THEMATICS returns competitive recall and precision with no cost for lack of similarity. THEMATICS thus holds advantages and is effective for novel folds, for engineered structures, and for proteins that do not possess a sufficient set of homologues to obtain meaningful conservation scores.

THEMATICS predicts the binding residues in apo structures, even when there is substantial change in the backbone conformation upon ligand binding and even when these binding residues are physically separated in the apo form. THEMATICS is effective because it utilizes the special electrostatic properties of active site residues. At these sites, there tends to be large interaction between protonation events on the ionizable species. These special electrostatic properties of the catalytic and binding residues are sufficiently preserved in the open form, although diminished compared to the closed form, such that the residues that constitute the two halves of the interaction site may be identified by statistical analysis in an open, unbound structure.

One of the difficulties in catalytic site prediction is that not all of the important residues are included in the annotated database and some have not been studied. Hence some of the "false positives" really are not false. Actual precision and MCC values for each method therefore are really higher and actual false positive rates are really lower than reported. The true quality and effectiveness of THEMATICS and the other top-performing methods is better than that reflected in the precision and MCC values; the relative performance of the different methods is also apparent from these values.

While computed protonation properties give excellent performance when used alone, they show tremendous potential in combination with other, complementary methods. A beta version of a simplified form of the THEMATICS site predictor is now available at: http://pfweb.chem.neu.edu/thematics/submit.html .

Acknowledgments

This work is supported by the National Science Foundation grant MCB-0517292.

References

[1]   Amitai, G., A. Shemesh, E. Sitbon, M. Shklar, D. Netanely, I. Venger, and S. Pietrokovski, Network analysis of protein structures identifies functional residues, *J Mol Biol*, 344:1135-1146, 2004.

[2]   Bartlett, G.J., C.T. Porter, N. Borkakoti, and J.M. Thornton, Analysis of Catalytic Residues in Enzyme Active Sites, *J Mol Biol*, 324:105-121, 2002.

[3]   Ben-Shimon, A. and Eisenstein, M., Looking at Enzymes from the Inside out: The Proximity of Catalytic Residues to the Molecular Centroid can be used for Detection of Active Sites and Enzyme-Ligand Interfaces, *Journal of Molecular Biology* 351:309-326, 2005.

[4]   Berman, H.M., Westbrook, J., Feng, Z., Gilliland, G., Bhat, T. N., Weissig, H., Shindyalov, I. N., Bourne, P. E., The Protein Data Bank, *Nucleic Acids Res*, 28:235-242, 2000.

[5]   Carugo, O., and S. Pongor, A normalized root-mean-square distance for comparing protein three-dimensional structures, *Protein Sci*, 10:1470-1473, 2001.

[6]   Cheng, G., B. Qian, R. Samudrala, and D. Baker, Improvement in protein functional site prediction by distinguishing structural and functional constraints on protein family evolution using computational design, *Nucleic Acids Res*, 33:5861-5867, 2005.

[7]   Clodfelter, K.H., Waxman, D.J. and Vajda, S., Computational Solvent Mapping Reveals the Importance of Local Conformational Changes for Broad Substrate Specificity in Mammalian Cytochromes P450, *Biochemistry*, 45:9393-9407, 2006.

[8]   Di Cera, E., and Z.-Q. Chen, The Binding capacity is a probability density function, *Biophys J*, 65:164-170, 1993.

[9]   Di Cera, E., S.J. Gill, and J. Wyman, Binding Capacity: Cooperativity and buffering in biopolymers, *Proc Natl Acad Sci U S A*, 85:449-452, 1988.

[10] Gilson, M.K., Multiple-site titration and molecular modeling: two rapid methods for computing energies and forces for ionizable groups in proteins, *Proteins*, 15:266-282, 1993.

[11] Gutteridge, A., G. Bartlett, and J.M. Thornton, Using a neural network and spatial clustering to predict the location of active sites in enzymes, *Journal of Molecular Biology*, 330:719-734, 2003.

[12] Ko, J., L.F. Murga, P. Andre, H. Yang, M.J. Ondrechen, R.J. Williams, A. Agunwamba, and D.E. Budil, Statistical Criteria for the Identification of Protein Active Sites Using Theoretical Microscopic Titration Curves, *Proteins: Structure Function Bioinformatics*, 59:183-195, 2005.

[13] Laurie, A.T.R., and R.M. Jackson, Q-SiteFinder: An energy-based method for the prediction of protein-ligand binding sites, *Bioinformatics*, 21:1908-1916, 2005.

[14] Madura, J.D., J.M. Briggs, R.C. Wade, M.E. Davis, B.A. Luty, A. Ilin, J. Antosiewicz, M.K. Gilson, B. Bagheri, L.R. Scott, & J.A. McCammon, Electrostatics and diffusion of molecules in solution - Simulations with the University of Houston Brownian Dynamics program, *Comp Phys Commun*, 91:57-95, 1995.

[15] Nimrod, G., F. Glaser, D. Steinberg, N. Ben-Tal, and T. Pupko, In silico identification of functional regions in proteins, *Bioinformatics*, 21 Suppl 1:i328-i337, 2005.

[16] Ondrechen, M.J., J.G. Clifton and D. Ringe, THEMATICS: A simple computational predictor of enzyme function from structure, *Proc. Natl. Acad. Sci. (USA)*, 98:12473-12478, 2001.

[17] Panchenko, A.R., Kondrashov, F. and Bryant, S., Prediction of functional sites by analysis of sequence and structure conservation, *Protein Sci*, 13:884-892, 2004.

[18] Petrova, N. and Wu, C., Prediction of catalytic residues using Support Vector Machine with selected protein sequence and structural properties, *BMC Bioinformatics*, 7:312, 2006.

[19] Porter, C.T., Bartlett, G.J. and Thornton, J.M., The Catalytic Site Atlas: a resource of catalytic sites and residues identified in enzymes using structural data, *Nucl. Acids Res.*, 32:D129-133, 2004.

[20] Silberstein, M., S. Dennis, L. Brown, T. Kortvelyesi, K. Clodfelter, and S. Vajda, Identification of substrate binding sites in enzymes by computational solvent mapping, *J Mol Biol*, 332:1095-1113, 2003.

[21] Wei, Y., J. Ko, L.F. Murga, and M.J. Ondrechen, Selective prediction of interaction sites in protein structures with THEMATICS, *BMC Bioinformatics*, 8:119, 2007.

[22] Wyman, J., Linked functions and reciprocal effects in hemoglobin: A second look, *Adv Protein Chem*, 19:223-286, 1964.

[23] Youn, E., B. Peters, P. Radivojac, and S. D. Mooney, Evaluation of features for catalytic residue prediction in novel folds, *Protein Sci*, 16:216-226, 2007.

AN ACCURATE AND EFFICIENT ALGORITHM FOR PEPTIDE AND PTM IDENTIFICATION BY TANDEM MASS SPECTROMETRY

KANG NING
ningkang@comp.nus.edu.sg

HOONG KEE NG
nghoongk@comp.nus.edu.sg

HON WAI LEONG
leonghw@comp.nus.edu.sg

Department of Computer Science, School of Computing, National University of Singapore, Computing 1, Singapore 117590

Peptide identification by tandem mass spectrometry (MS/MS) is one of the most important problems in proteomics. Recent advances in high throughput MS/MS experiments result in huge amount of spectra. Unfortunately, identification of these spectra is relatively slow, and the accuracies of current algorithms are not high with the presence of noises and post-translational modifications (PTMs). In this paper, we strive to achieve high accuracy and efficiency for peptide identification problem, with special concern on identification of peptides with PTMs. This paper expands our previous work on PepSOM with the introduction of two accurate modified scoring functions: S_λ for peptide identification and S_λ^* for identification of peptides with PTMs. Experiments showed that our algorithm is both fast and accurate for peptide identification. Experiments on spectra with simulated and real PTMs confirmed that our algorithm is accurate for identifying PTMs.

1. Introduction

Peptide identification by tandem mass spectrometry (MS/MS) is a very challenging problem that receives wide attention by computational biologists. Huge amount of MS/MS spectra generated requires the most taxing of resources to process. With the presence of noises and post-translational modifications (PTMs) that further complicate the problem, current algorithms for peptide identification are not very accurate.

Approaches for peptide identification can be categorized into database search algorithms [1-3] and *de novo* algorithms [4-7]. The former return peptide sequences that match the parent mass of the experimental spectrum via some scoring functions. Their accuracies largely depend on the completeness of the database, and the process is usually slow. Additionally, they generally do not perform well for peptides sequences not already known as well as peptides with PTMs. On the other hand, *de novo* algorithms interpret peptide sequences from spectrum data purely by analyzing the intensity and correlation of the peaks in the spectrum data. They can retrieve tags from spectrum with high accuracy [3], and the process is very fast (always within a minute). However, their performance quickly deteriorates in the presence of noises and PTMs.

Striving for high efficiency and identification accuracy for peptide identification by MS/MS with the presence of PTMs is an essential issue, which is especially important for experts in the analysis of results in the "wet laboratory". This paper focuses on this issue.

Recently, the InsPecT algorithm [8] was proposed, which first generates a set of highly accurate tags from spectrum, and then use these tags to filter peptide sequences in database. As *de novo* is imperfect, multiple tags are produced for each spectrum. The accuracy of InsPecT depends on the quality of the tags but even in the context of up to a dozen modifications, it performs reasonably well. Another interesting aspect of InsPecT is that it uses automata to search for peptide sequences. For a batch of spectrum data, the process can be very quick (about 10 ms per spectrum).

For algorithms based on tags for coarse filtering, though using tags can achieve reasonable efficiency, the quality of candidates is very dependent on the quality of tags which in turn are highly dependent on the quality of the spectra.

Identification of the PTMs is another important problem. Most of current algorithms that are able to identify PTMS [1-3] first define a set of possible PTMs before peptide identification. This approach is limited by the number of predefined PTMs. The InsPecT algorithm takes another approach, which allows blind PTM search (no specified PTM). This approach can discovery virtually every possible PTMs in the peptides, but may has high false positive rate.

Previously, we proposed the PepSOM algorithm [9] which can achieve high efficiency for peptide identification by database search based on SOM and MPRQ. However, the accuracies of the PepSOM results are not very satisfactory. This is because after candidate peptides are retrieved from database, they are scored and ranked by SPC, which is not an accurate scoring function especially on noisy spectra and spectra with PTMs. Apparently, comparing candidate peptides with experimental spectrum alone is not enough, so in this paper candidate peptides are also compared with highly-reliable tags generated from experimental spectrum by our *de novo* algorithm.

2.   Computational Model and Algorithm

In this section, we briefly formulate the problem of peptide identification by MS/MS, and describe a *de novo* algorithm to generate multi-charge strong tags, and mention PepSOM. Due to space constraints, refer to [9-11] for more details. Then we introduce the scoring functions to score and rank candidate peptides (with PTMs).

Problem Formulation and Multi-charge Strong Tags

To introduce multi-charge strong tags, we first define some general terms. In tandem mass spectrometry, a peptide sequence ρ will be fragmented into a spectrum S, which is composed of many peaks $\{p_1, p_2, \ldots, p_n\}$. Each of the peaks p_i is represented by its *intensity*(p_i) and mass-to-charge ratio $mz(p_i)$. If peak p_i is not noise, then it will represent a fragment ion of ρ. We say that peak p_i is a *support peak* for the fragment q and we say that the fragment q is supported by the peak p_i. A peak p_j is a support peak for the peak p_i if both of them are support peaks for the same fragment q.

In the problem of peptide identification by MS/MS, the input is the mass spectrum S, and the output is the putative peptide sequence P from which the spectrum is generated. The theoretical spectrum completely characterizes all possible peaks for a peptide by considering all ion types. On the contrary, experimental spectrum seldom completely characterizes all possible peaks for peptides, and it contains many noises.

Recently, we proposed the GST-SPC algorithm [11] which was shown to generate high quality tags (called *multi-charge strong tags*, or simply *tags*). In the first phase, GST-SPC computes a set of all tags. Then GST-SPC tries to link these tags by their mass differences, and computes a peptide sequence that is optimal with respect to shared peaks count (SPC) from all peptides derived from tags. Since previous results show that the tags generated by GST-SPC are accurate, in this paper we use multi-charge strong tags generated by the first phase of GST-SPC in scoring the candidate peptides.

Binning of Peaks

Binning is performed to convert peptides (transformed to theoretical spectra) in database to high-dimensional vectors in vector space.

A spectrum is divided into fixed intervals by mass-to-charge ratios; within each interval the peak with the highest intensity is chosen. To further improve the performance of binning, we incorporated noise removal and scoring of bins after binning. For the full details, refer to [9]. Proper values of tolerance used in binning can preserve accuracies, while decreasing the computational cost greatly, especially for noisy spectra. For the ion trap datasets in this paper, mass tolerance $m_t{}^*$ is set to be 0.5 Da and the mass range of bin m_{bin} is set to be 0.25 Da. With the process of binning and noise removal, only those significant bins (peaks) are kept, resulting in better accuracy and efficiency.

SOM and Multi-Point Range Query

Self-organizing map (SOM) [12] is used to transform high-dimensional vectors to 2D points on a plane. In our algorithm, spectrum similarity could be transformed to vector similarity and then to 2D metric similarity. Subsequently, MPRQ [13, 14] is used for multi-point similarity query on the plane to efficiently identify candidates.

For peptide identification, once the theoretical spectra for the peptide sequences in the database are mapped as 2D points on a plane by SOM, we transform the query (experimental) spectra into query points in plane and proceed to query. MPRQ algorithm also accepts as input a parameter d that controls the radius of the search distance. The larger the value of d, the more candidate peptides will be returned. MPRQ can efficiently process multiple input points *simultaneously* during query, effectively performing configurable multi-spectra similarity search on database of known peptides.

Scoring and Ranking

To achieve high accuracy in peptide identification, the most important step is the scoring and ranking of candidate peptides results from database search. We had shown in [9] that by using SPC alone for scoring of candidate peptides results in low identification accuracy. Therefore, here we also compared the candidate peptides with tags generated by GST-SPC. This approach combines the comparison of candidate peptides with experimental spectrum and also with tags.

We now introduce *SPC score* and S_{tag} *score*. SPC score is computed as the number of peaks of the same mass-to-charge ratios (within tolerance) between experimental and theoretical spectrum of the candidate peptide, over the number of peaks in experimental spectrum. The S_{tag} score, which measures the similarity of candidate peptide to tags, is computed as the ratio of candidate peptide that can match one or more tags at the correct position (within the range of [0,100] Da), over the length of the candidate peptide. For example, given the candidate peptide "VAQLEQVYIR" and two tags "VAK" and "IVYLR" starting from mass of 0 Da and 550 Da, respectively. If we do not allow mismatch, then S_{tag} is computed as (3+4)/10=0.7; if we allow up to one mismatch, then S_{tag} is computed as (3+5)/10=0.8. To score and rank candidate peptides, we define and use a scoring function S_λ which is a weighted sum of SPC and S_{tag} scores.

$$S_\lambda = w_1 \cdot SPC + w_2 \cdot S_{tag} \tag{1}$$

The weights are derived empirically. We selected a large amount of (experimental spectrum, peptide) pairs with high confidence (Xcorr $\geq$ 2.5) from the ISB dataset (details in Table 1). We then compute their SPC and S_{tag} scores and tried different combinations of these two scores. We found that $w_1 = 0.1$ and $w_2 = 0.9$ give discriminative results (details omitted due to space limit), and the results were normalized this way.

For PTM identification, it is observed that because of peptide fragmentation such as loss of water and ammonia, PTMs such as phosphorylation, as well as the errors introduced by the mass spectrometer ion detector, *mass shifts* in spectra are very common. Specifically, each PTM corresponds to a set of shifted peaks in experimental spectrum. And highly possible PTMs should have strong support represented by such a set of mass shifts. Here, we use a *modified SPC scoring function* (SPC*) that can better handle sets of mass shifts in spectra for identification of peptides with PTMs.

To illustrate the mass shifts by PTMs, Fig. 1 shows an example of an experimental spectrum which is identified to be $I^{+43}TFYEDR$ (with PTM) by [15]. We compared it with theoretical spectra (with Δ^R) for two peptides, $I^{+43}TFYEDR$ and ITFYEDR. The intensity of the peaks in theoretical spectrum in not known, and we assume b-ion and y-ion peaks to have higher intensity than peaks of other ion types.

Comparison between theoretical spectrum of I^{+43}TFYEDR and that of ITFYEDR clearly shows a set of peak shifts, corresponding to the PTM on amino acid "I". It is hard to see from Fig. 1 which theoretical spectrum is similar to experimental spectrum without computation of SPC score. The SPC score for I^{+43}TFYEDR is 0.287, while that for ITFYEDR is 0.154. A big difference is observed between them. Using the above spectrum example, we performed database search, and the top candidate peptides are ITFYEDR and LTFYEEV. The tag generated by GST-SPC algorithm is "TFYED". We computed the S_λ score for these two peptides as $S_\lambda = 0.1*0.154 + 0.9*0.71 = 0.654$ for ITFYEDR, and $S_\lambda = 0.1*0.021 + 0.9*0.57 = 0.515$ for LTFYEEV.

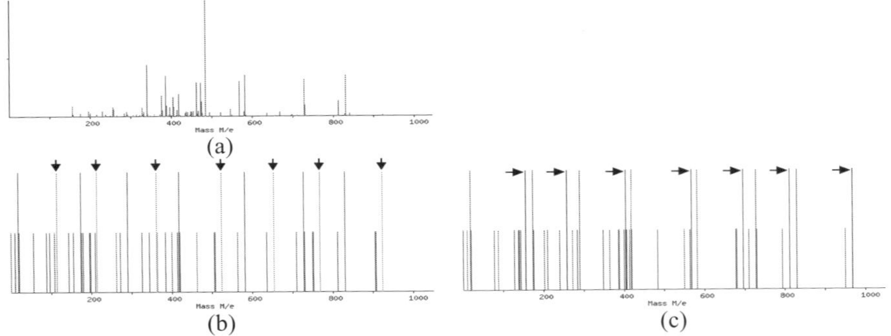

Fig. 1. Example of mass shift caused by PTMs. (a) Experimental spectrum for I^{+43}TFYEDR, (b) Theoretical spectrum for ITFYEDR, (c) Theoretical spectrum for I^{+43}TFYEDR. The shifted b-ion peaks are indicated with arrows in (b) and (c).

At each cleavage site, we assume each of $i*m_{bin}$ Da for all $-60 \le i*m_{bin} \le 60$ (60 Da was determined empirically) as a putative mass shift. We define $SPC_{i,j}$ as the SPC score between experimental and theoretical spectrum of candidate peptide P, where we assume a mass shift of $i*m_{bin}$ Da at cleavage site j of P. It is easy to see that $SPC_{0,j}$ is the SPC score of experimental and theoretical spectrum without mass shift at cleavage site j. If the largest $SPC_{i,j}$ for cleavage site j is obtained with $i > 0$, then this cleavage site j is a putative PTM site with mass shift of $i*m_{bin}$ Da, with the *PTM score* of

$$S_{PTM}(j) = (SPC_{i,j} - SPC_{0,j}) \tag{2}$$

If $S_{PTM}(j)$ is greater than a threshold T_{PTM} (determined empirically to be 0.1), then we say that this putative PTM site is significant, and we identify this as a PTM in the peptide. In $S_{PTM}(j)$ function, a series of mass shifts introduced by a single PTM is regarded as a whole event, which is more realistic.

Take the same example above, we computed the S_{PTM} score at cleavage sites for I^{+43}TFYEDR. Since "TFYED" is a tag, only the cleavage sites after "I" ($j = 1$) and before "R" ($j = 5$) were analyzed. Results are that $S_{PTM}(1) = 0.133$ with mass shift of 43 Da, and $S_{PTM}(5) = 0$ with mass shift of 0. Since $S_{PTM}(1)$ is above T_{PTM} while $S_{PTM}(5)$ is below it, we thus identify this PTM correctly.

We further define $SPC_{\{i_1..i_q\},\{j_1..j_q\}}$ as the SPC score between experimental and theoretical spectrum of identified peptide P, where a mass shift of $\{i_1*m_{bin}...i_q*m_{bin}\}$ Da matches with cleavage site $\{j_1...j_q\}$ of P, in which each $S_{PTM}(j)$ is greater than T_{PTM}. The corresponding $S_{PTM}*$ is defined as

$$S_{PTM}* = \sum_{j=1}^{K} (SPC_{\{i_1..i_q\},\{j_1..j_q\}} - SPC_{\{i_1..i_q\},\{0..0\}}) \mid (S_{PTM}(j) > T_{PTM}) \tag{3}$$

in which K is the length of the peptide. $S_{PTM}*$ indicates the significance of PTMs in a spectrum. Though we have considered multiple PTMs in $SPC_{\{j_1..j_q\}}$, experiments show that there is usually not more than one PTM per spectrum. Taking PTMs into consideration, the modified S_λ score is then defined as

$$S_\lambda* = w_1 \cdot SPC_{\{i_1..i_q\},\{j_1..j_q\}} + w_2 \cdot S_{tag} \qquad (4)$$

which can be used for identification of peptides with PTMs. The weight w_1 and w_2 are again determined empirically. To derive the weights, we selected a large amount of (experimental spectrum, peptide) pairs of peptides with PTMs of high confidence (*p*-value of 0.05 or better, computed the same way as in InsPecT [3]) from ISB datasets. We then computed their $SPC_{\{i_1..i_q\},\{j_1..j_q\}}$ and S_{tag} scores and tried many different combinations of these two scores. Similar to that used for S_λ scoring function, the results (details not shown) indicate that $w_1 = 0.3$ and $w_2 = 0.7$ give discriminative results.

Take the same example as illustrated above; we have computed the $S_\lambda*$ score for the spectrum against the two candidate peptides ITFYEDR and LTFYEEV. Results show that ITFYEDR has score $S_\lambda* = 0.3*0.154 + 0.7*0.71 = 0.543$ while I^{+43}TFYEDR has score $S_\lambda* = 0.3*0.287 + 0.7*0.71 = 0.583$; and LTFYEEV has score $S_\lambda* = 0.3*0.021 + 0.7*0.57 = 0.405$ while L^{+43}TFYEEV (the best PTM identified on sequence LTFYEEV) has score $S_\lambda* = 0.3*0.144 + 0.7*0.57 = 0.442$. These results show that I^{+43}TFYEDR has the best $S_\lambda*$ score, indicating that $S_\lambda*$ score is discriminative.

Apparently, $S_\lambda*$ is much more expensive in terms of computation than S_λ. However, after coarse filtering, we only need to consider a small number of candidate peptides; computing $S_\lambda*$ on these limited set of candidate peptides is still acceptable.

Our Algorithm

In this paper, we use a peptide identification algorithm that is a combination of database search technique and *de novo* technique. It has the following steps: (i) both peptides in database and experimental spectra are first converted to high-dimensional vectors via binning; (ii) the vectors are mapped to 2D plane with SOM; (iii) candidate peptides are then selected from database with MPRQ; and (iv) these candidate peptides are scored and ranked (fine filtered) by a scoring function that compares them with the experimental spectrum as well as multi-charge strong tags generated by a *de novo* algorithm [11]. Steps (i)-(iii) are coarse filtering steps, in which spectra similarity is transformed to vector similarity and then to 2D points metric distance similarity. These steps are similar to those in PepSOM [9]. Step (iv) is a fine filtering step in which the candidate peptides are scored and ranked by comparing them with experimental spectrum and tags generated by the GST-SPC algorithm. At the end of step (iii), if we assume that there is no PTM in the spectrum and want to perform fast peptide identification, then S_λ scoring function is used. Otherwise $S_\lambda*$ scoring function is used for identification of peptides with PTMs.

3. Experiments

Experiments were performed on a 3.0 GHz PC with 1.0 GB main memory running Linux. Our algorithm is implemented in C++ and Perl. SOM_PAK [16] was the SOM implementation used. For analysis and comparison, we had selected established algorithms with freely available software: two database search algorithms, Sequest [1] and InsPecT [3]; as well as two *de novo* algorithms, Lutefisk [7] and PepNovo [5].

Spectrum datasets (query datasets) were obtained from Open Proteomics Database [17], PeptideAtlas [18] and Institute for Systems Biology [19]. All of the experimental mass spectra were ion trap data having low mass resolution. As the statistical evaluation of the correlation of spectrum and peptide is still a difficult open problem, we treated

Sequest result with Xcorr ≥ 2.5 as ground truth, which is considered reliable.

The PeptideAtlas spectrum dataset A8_IP were obtained from Human Erythroleukemia K562 cell line. Electrospray ionization source of an LCQ Classic ion trap mass spectrometer (ThermoElectron, San Jose, CA) was used, and DTA files were generated from MS/MS spectra using TurboSequest. All 44 spectra that were identified with Xcorr ≥ 2.5 were chosen. For OPD, the spectrum dataset used was opd00001_ECOLI, *Escherichia coli* spectra 021112.EcoliSol 37.1(000). The spectra were obtained from *E. coli* HMS 174 (DE3) cell, which is grown in LB medium until ~0.6 abs (OD 600). The spectra were generated by the ThermoFinnigan ESI-Ion Trap "Dexa XP Plus" and the sequences for these spectra were validated by Sequest. The ISB dataset was generated using an ESI source from a mixture of 18 proteins, obtained from ion trap mass spectrometry, and consists of spectra of up to charge 3. Most importantly, these ISB datasets were annotated by a few algorithms [8, 15] to be free of PTMs (refer to http://www.systemsbiology.org/extra/protein_mixture.html).

The databases that we have used contain peptides from the respective protein sequences dataset. Specifically, *E. coli* K12 protein sequences for OPD datasets, IPI HUMAN protein sequences for PeptideAtlas dataset and human plus control protein mixture for ISB dataset were used. As the number of protein sequences were very large for PeptideAtlas (60,090) and ISB (88,374) datasets, we used only the protein sequences corresponding to spectra identified with Xcorr ≥ 2.5 (our ground truth). However, the sizes of databases were still very large because of many peptides. The parameters for the generation of databases, the query datasets and theoretical spectra are shown in Table 1.

Table 1. Parameters for the generation of databases and theoretical spectra.

Parameters	Values		
	PeptideAtlas	OPD	ISB
No. of protein sequences	31	4,279	3,553
Total database size	9,421	494,049	1,248,212
Query size	44	202	995
Fragments mass tolerance	0.5 Da		
Parent mass tolerance	1.0 Da		
Modifications	–		
Charge	+2, +3		
Ion type	a, b, y, $-H_2O$, $-NH_3$		
Missed cleavages	0		
Protease	Trypsin		
Mass range	0-5000 Da		

To compare the different algorithms, the following accuracy measures were used:

$$Recall = \frac{\#\,correct}{|\rho|} \qquad (5)$$

$$Precision = \frac{\#\,correct}{|P|} \qquad (6)$$

where *#correct* is the *number of correctly identified amino acids*. For two amino acids in the correct peptide ρ and the respective identification result P, only if their positions do not have a difference of more than 100 Da (determined empirically) and they are of the same amino acids (except (I, L), as well as (K, Q)), do they contribute one count to *#correct*. A high *Recall* being that the algorithm recovers a large portion of the correct peptide. For a fair comparison with algorithms like PepNovo that only outputs the highest scoring tags (subsequences), we also use a *Precision* measure, which measures how many of the results are correct. Note that these recall and precision measures are different from sensitivity and specificity measures used in PepSOM paper, since there is a position constraint on amino acids in recall and precision measures, rather than only using LCS to measure *#correct* in sensitivity and specificity in PepSOM.

Experiments on Peptide Identification

In this subsection, we performed experiments using peptides without PTMs. Firstly, we analyzed the quality of the tags generated by the GST-SPC algorithm. We measured the ratio of completely correct tags in the results, as well as recall and precision of the tags. Results are shown in Table 2. Note that we had only analyzed the quality of tags on ISB spectra in our previous study [11]. By also measuring OPD and PeptideAtlas datasets, we empirically proved the accuracy of tags on a variety of datasets.

Table 2. Statistical results on the quality of the generated tags. "No. of tags per spectrum" shows the average number of tags generated per spectrum. "No. of complete correct per spectrum" measures the average number of tags identified that are completely correct (i.e. identified with 100% precision). "Complete correct accuracy" is the ratio of "completely correct tags" to number of tags on average. The recall and precision results are obtained from tags by the GST-SPC algorithm.

Datasets	Query Size	Average Peptide length	No. of Tags per Spectrum	No. of Complete Correct per Spectrum	Complete Correct Accuracy	Recall	Precision
OPD	202	10.14	7.42	6.01	0.81	0.43	0.43
PeptideAtlas	44	10.02	9.76	6.83	0.70	0.40	0.36
ISB	995	19.37	6.19	4.61	0.74	0.36	0.32

From Table 2, we observed that more than 1/3 of the amino acids in real peptide sequences (recall) can be correctly identified by tags. Also, when the tags are generated, more than 70% of the tags are completely correct, meaning that the tags generated are reliable. Since each tag is at least one amino acid in length, it can also be observed that a significant amount of tags are overlapping. For more reliable results in the following experiments, only non-overlap tags with high scores (determined by GST-SPC) are used.

Secondly, we investigate the quality of candidate peptides identified by MPRQ and SOM. We analyzed the search distance d on the accuracy of search results on datasets of different sizes. Note that similar spectra that overlap on the same 2D point can be losslessly retrieved by our algorithm since it has built an index for these overlapping spectra. The candidate peptides are scored and ranked by SPC score only. First-rank peptide represents the peptide with theoretical spectrum that has the highest SPC score against the experimental spectra. Best-match peptide is the peptide among all candidates that match with the "real" peptide with the highest precision (recall).

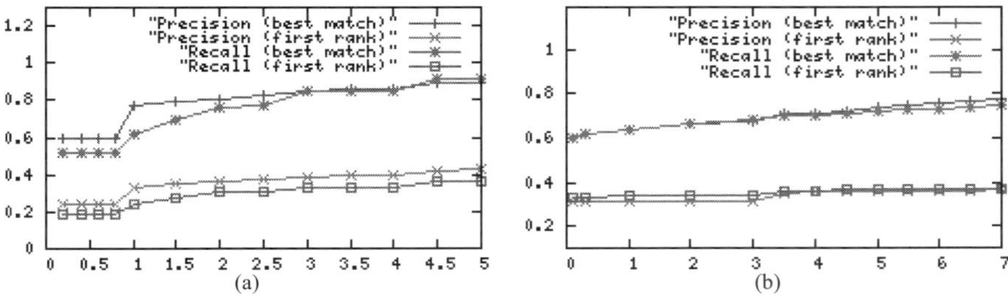

Fig. 2. The effects of increasing distance d on recall and precision on (a) PeptideAtlas dataset (b) ISB dataset.

Analysis of the search distance d (Fig. 2) show that the recall and precision of best-match peptides are much higher that those for first-rank peptides, indicating that (i) SPC score alone is not a good scoring function; and (ii) a properly designed scoring function can improve the identification accuracies significantly. The PeptideAtlas, OPD and ISB datasets are datasets of increasing sizes. Results also indicate that for larger datasets, the search distance should also be bigger to achieve high recall and precision. Therefore, for datasets of large sizes, we used larger search distances. Specificly, we used d=1.0 for PeptideAtlas datasets, d=2.5 for OPD datasets and d=3.5 for ISB datasets. Fig. 2(b) also shows that as search distance d increases, more candidate peptides are returned increasing both recall and precision. However, when d grows beyond the similarity clusters in the

SOM, not so relevant candidates returned bring down the recall and precision values.

The average search time per spectrum is less than 11 ms. This is comparable to InsPecT (which average 10 ms per spectrum with default settings, but based on smaller database), one of the fastest database search algorithms.

Another important question is, among the candidate sequences, what is the ratio of them being identical to the real peptide sequences. We have found hat when we consider all of the candidates, the "complete correct accuracy" is much higher; for OPD dataset it is 69.5%, PeptideAtlas 63.1% and ISB 65.3%. And if we allow up to two amino acids difference from real peptide sequences, the ratios increase to 80.1%, 85.3% and 78.6% respectively. Therefore, given a good scoring function, the peptide identification accuracy can be significantly increased. As the size of the candidate sequences generated by our algorithm is rather small (refer Table 6), we believe these high ratios indicate good performance of the SOM and MPRQ as coarse filtering.

Table 3. Comparison of different algorithms on the accuracies of peptide identification. In each column, the "Precision / Recall" values are listed.

Datasets	Database Size	Query Size	InsPecT	Lutefisk	PepNovo	Our Algorithm
OPD	494,049	202	0.580 / 0.542	0.101 / 0.006	0.232 / 0.186	**0.582 / 0.603**
PeptideAtlas	9,421	44	0.801 / 0.389	0.149 / 0.057	0.275 / 0.128	**0.521 / 0.457**
ISB	1,248,212	995	0.584 / 0.621	0.011 / 0.022	0.548 / 0.561	**0.594 / 0.695**

Subsequently, we compared our algorithm with other algorithms. For our algorithm, S_λ scoring function is used, and the results are based on peptides with the best score. The results with first rank given by these algorithms were used for analysis.

We observe from Table 3 that both precision and recall of our algorithm are better than Lutefisk and PepNovo (both *de novo* algorithms). This is reasonable since *de novo* algorithms do not utilize any information from databases. But even when comparing their results with the quality of tags generated by our algorithm (Table 2), we notice that the quality of tags generated by our algorithm is better than peptide identification results by Lutefisk, and comparable with that by PepNovo. Although InsPecT has higher precision, our results outperform InsPecT in recall. Specifically, for the OPD dataset, both the algorithms have precision of about 0.58, but our algorithm has higher recall. For the PeptideAtlas dataset, the precision of our algorithm is much worse than that of InsPecT, but the recall is 17% better. For the ISB dataset, both InsPecT and our algorithm have similar precision, but recall of our algorithm is higher. This means that our algorithm can identify more portion of the real peptide.

We have also observed that by scoring peptide candidates using S_λ, both precision and recall consistently increase (last column of Table 3), compared with only using SPC score (Fig. 2). This proves the superiority of S_λ scoring function.

Experiments on PTM Identification

PTM identification is of great importance to current mass spectrum analysis. To analyze PTMs, we first performed experiments on experimental spectra *in silico* with artificially added PTMs (we call these simulated PTMs). We selected spectra from ISB datasets that are annotated to be free of PTMs. For every peptide, the PTM that we had artificially added is phosphorylation for every amino acids involved. In the corresponding experimental spectrum, we shifted every peak that corresponds to the respective peptide fragment according to the restricted ion types Δ^R. Summary of modifications:

Modification	Amino acid involved	Context	Mass difference (Da)
Phosphorylation	T,S,Y	PTM	+79.97

Our algorithm is not designed specifically for phosphorylation per se, but can also be easily applied to detect other types of PTMs. These can be shown in experiments on the detection of PTMs on real datasets, using ISB spectra [19] that contain PTMs but are distinct from the "ISB dataset" we used above which do not. It was found that there are PTMs in these ISB datasets [15], and their identifications (called UCSD annotation) are found at http://www.systemsbiology.org/extra/UCSD_supplemental_identifications.txt. There are 551 spectra with at least one PTM identified by InsPecT from a total of 2,799 ISB spectra. The results of the "UCSD annotation" were treated as ground truth, since they are annotated by InsPecT with an annotation of p-value 0.05 or better, indicating reliability.

The UCSD datasets contains those ISB spectra with PTMs identified by InsPecT. However, for analysis and comparison purpose, we have also applied different algorithms on other ISB datasets for possible identification of new PTMs in these spectra. ISB spectra that are different from previously described datasets were selected; we refer to this dataset and our annotations as "NN annotation" dataset. This dataset contains 3,000 spectra. Again, we treat PTMs identified with p-value 0.05 (computed the same way as in InsPecT) or better as ground truth.

S_λ^* was used to identify peptides with PTMs. Peptide identification accuracy is measured as the percentage of candidate peptides that contain the exact original (unmodified) peptide. PTM identification accuracy is measured as the percentage of results in which the best-score PTM (Eq. 3) identification is *correct*, where PTM identification is defined as correct if (i) the original peptide is identified correctly *and* (ii) the PTM site and the value of mass shift are identified correctly. For instance, a peptide (with PTM) "AS^{+80}RK" is identified correctly, if "ASRK" is identified correctly, *and* the PTM site and PTM mass shift (+80 Da) after "S" are identified correctly.

Firstly, we have analyzed the accuracies of PTM identification on simulated PTMs. We used tags of specific lengths for analysis. The results on ISB spectra with simulated PTMs are shown in Table 4.

Table 4. Accuracies (%) of PTM identification from simulated spectra by tags of different lengths. The columns with Top $k = 1, 2, 3, 4$ represent the (peptide / PTM) identification accuracies for top-k. "No limit" means that the best-score tags are used without any length limit. "Filtration ratio" is computed as the number of candidates after tag filtration over the number of candidates after MPRQ. "Time" is the total time to identify the peptides and PTMs for 995 spectra. Results without using tags are also shown.

Database Size	Query Size	Tag length	Top 1	Top 2	Top 3	Top 4	All	Filtration Ratio	Time (s)
1,248,212	995	3	46.7 / 30.2	50.1 / 36.3	62.6 / 40.5	69.2 / 46.5	**71.3 / 60.1**	0.0148	65.6
		4	56.9 / 34.6	40.5 / 25.6	44.4 / 32.6	51.0 / 39.0	**63.3 / 50.0**	0.0021	67.5
		No limit	46.8 / 32.9	52.0 / 36.1	58.3 / 43.3	64.4 / 50.1	**72.8 / 59.1**	0.0491	66.6
		No tag	31.7 / 26.4	35.5 / 26.6	41.1 / 35.2	46.9 / 39.5	**56.7 / 40.8**	–	70.7

From the results above, it can be observed that sequence tags of length 3 and 4 are able to further filter out candidates from the results of SOM and MPRQ. With reduced candidates, the accuracy for PTM identification increased. Compared with results without tags, the percentages of search results that contain the exact correct peptide are significantly higher. For example, for filtration with tags of length 3, about 46.7% to 71.3% of original peptides are identified correctly. Increase filtration tags length to 4 decreases peptide identification accuracies, but using best-score tags without any length limit do not show such decrease. PTM identification accuracies show similar patterns. These indicate that although longer tags may have lower recall, the best-score tags are of high recall, regardless of their length.

We notice that the process time based on tags of length 4 is greater than that on tags of length 3. We think that though longer tags may filter out more candidates, which makes later scoring faster, the filtering step itself is more time consuming than those based on tags of length 3, so that the total time is longer. Also, we notice that the processing time of tags without length limit is shorter than that of the processing time of tags with length 4. Since we have observed that the best-score tags are of average length

> 4 (details not shown), this indicates that the best-score tags (without any length limit) can filter out even more candidates, which makes the scoring step faster.

The InsPecT algorithm (with blind PTM search) is also applied on these spectra with simulated PTMs. Results show that both the peptide and PTM identification accuracies are not as high as our algorithm. In all of the results (10 identifications per spectrum) given by InsPecT, the peptide identification accuracy is around 50%, while the PTM identification accuracy is approximately 33%.

We observed that by comparing candidate peptides with tags, a large ratio of candidate peptides that do not match with any tags will be filtered. We have also observed that the filtration ratio is small. For instance, the filtration ratio for tags with length 3 is 0.0148; for length 4 is 0.0021. This indicates that tags can further reduce the number of candidate peptides for further careful examination by S_λ^*.

Experiments on the identification of PTMs on real ISB spectra with "UCSD annotation" were also performed (Fig. 3(a)). Since experiments on simulated PTMs (Table 4) show that best-score tags with no length limits have the best accuracies, we used them here. Again, we treated PTMs identified with p-value 0.05 or better as ground truth. Results show that the filtration ratio of our algorithm is 0.062. The peptide identification accuracies are 42.0%, 45.7%, 48.2%, 50.6% and 55.5% for Top 1, 2, 3, 4 and all candidates, respectively; and the PTM identification accuracies are 31.6%, 33.1%, 34.8%, 40.2% and 41.8% for Top 1, 2, 3, 4 and all candidates, respectively. These values are slightly smaller than those on simulated spectra, and we think this is due to the different PTM types in real spectrum.

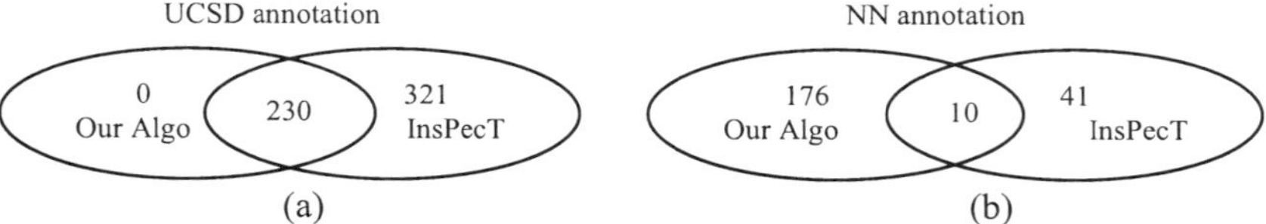

Fig. 3. The number of PTMs identified by our algorithm (no length limit on tags, Top 1 result) and InsPecT on "UCSD annotation" and "NN annotation" datasets. (a) There are 230 PTMs that both algorithms identified on "UCSD annotation". (b) There are 10 PTMs that both algorithms identified on "NN annotation".

Apart from the "UCSD annotations" on 2,799 spectra, we have also examined other ISB spectra in "NN annotation" on 3,000 spectra (Fig. 3(b)). Again, note that the UCSD datasets contains those ISB spectra with PTMs identified by InsPecT. On the other hand, the NN datasets contains spectra that are not subject to PTM identification by any algorithms before. We applied our algorithm and InsPecT algorithm (with blind PTM search) on these spectra; PTMs identification with p-value 0.05 or better are treated as ground truth. On these "NN annotation" spectra, InsPecT has identified 51 PTMs (among 78 peptide identifications with or without PTMs of p-value 0.05 or better) while our algorithm (using tags with no length limit) has identified 186 PTMs. Among those identified PTMs, 10 are identified by both algorithms. Interestingly, though InsPecT algorithm has identified 321 more PTMs in the UCSD annotation, in the NN annotation, 135 more peptides were identified using our algorithm. We think that this is because of the variance of the "UCSD annotation" and "NN annotation" datasets. Even though "NN annotation" dataset is selected randomly from ISB dataset, the complex nature of the ISB data itself makes such a big difference.

Listed below are some novel PTMs annotated (predicted) by our algorithm with high PTM scores (and low p-values) in "NN annotation" ISB spectra. Some of these annotations have experiment support (see References column), others are completely novel annotations. A full list of these novel PTM annotations will be provided upon request.

Table 5. Some representative novel PTMs identified by our algorithm.

Spectra (sergei_digest)	Peptides with PTMs	S_{PTM}	p-value	Notes and References
A_full_01.3541.3553.2	NFYFQCFNSG^{-37}LDSVLIADVPIEES	0.254	0.0207	
A_full_02.2185.2189.3	G^{-40}IIWGEDTLMEYLENPKK	0.197	0.0148	
A_full_03.3951.3951.2	DVPDARKC^{+53}ACASHVAKVA	0.563	0.0059	Also annotated by Unimod [20] without verification
A_full_05.1020.1020.2	LLKF^{+1}GQEV	0.349	0.0001	[21]
A_full_07.1737.1741.3	A^{-16}TAQADVVMMETPDELQAAVWEK	0.197	0.0062	
B_full_03.0832.0834.2	NALS^{+8}GNQNLEVWQLRLY	0.333	0.0007	

Efficiency

One of the most important aspects of our algorithm is that it is very fast. Table 6, reproduced from [9], explains. The coarse filtering rate is very low as we only need to compare each spectrum against the candidate peptides identified by MPRQ. Compared to the tandem cosine coarse filter used in [22] that filters to around 0.5% of the database, our algorithm has a better filtering efficiency.

Table 6. Candidates size, average candidate size and coarse filtering rate. "Candidates size" is the combined total results from coarse filtering of the database using the query size as input query points for the MPRQ algorithm. "Average Candidate Size" is the average peptide sequence candidates for each spectrum (query). "Coarse Filtering Rate" is computed by "average candidate size" over the database size.

Database	Database Size (peptides)	Query Size	Candidates Size	Average Candidate Size	Coarse Filtering Rate
OPD	494,049	202	68,610	339.7	0.069%
PeptideAtlas	9,421	44	654	14.9	0.158%
ISB	1,248,212	995	101,443	102.0	0.008%

After database search, the scoring of candidate peptides by S_λ scoring function is approximately 5 seconds per spectrum, while S_λ^* scoring function needs about 20 seconds for each spectrum. As comparison, for InsPecT the running time of blind search of PTMs is approximately 1 second per spectrum per megabyte of database (as stated in InsPecT documentation, and verified by our experiments). Relatively, our algorithm is very efficient on PTM identifications.

The program for our algorithm is available upon request.

4.   Conclusion and Future Work

This paper focused solely on the peptide identification problem, striving to achieve high identification accuracy and efficiency for peptide identification, especially for peptides with PTMs. An algorithm that transforms spectra similarity to similarity of vectors, and then to metric similarity (distance) of 2D points on a plane was used. The vectors are input to SOM to produce an indexable map in which MPRQ could use to find candidate peptides efficiently. Candidate peptides are fine-filtered with proposed scoring functions (S_λ for peptide identification and S_λ^* for identification of peptides with PTMs), which compare each of them with experimental spectrum and highly reliable tags generated by our GST-SPC algorithm.

Experiments lent strong support to the fact that by using S_λ scoring function that take into consideration score based on tags, the accuracies (precision and recall of the results) of our algorithm are high, yet still maintaining efficiency, especially for large batch processes. By using S_λ^* scoring function that take into consideration of mass shifts caused by PTMs, our algorithm can accurately identify peptides with PTMs. The novel PTMs that are predicted by our algorithm with high scores are interesting for manual verifications later in wet laboratories.

Recently, we noticed an algorithm Popitam (http://www.expasy.org/tools/popitam) that has similar scheme as ours. In Popitam, the scoring function is based on genetic programming (machine learning), which are quite different from our scoring function. Comparison of the two algorithms may be of interest in the future.

References

[1] J. K. Eng, A. L. McCormack, and I. John R. Yates, "An approach to correlate tandem mass spectral data of peptides with amino acid sequences in a protein database," *Journal of the American Society for Mass Spectrometry*, vol. 5, pp. 976-989, 1994.

[2] D. N. Perkins, D. J. C. Pappin, D. M. Creasy, et al., "Probability-based protein identification by searching sequence databases using mass spectrometry data," *Electrophoresis*, vol. 20, pp. 3551-3567, 1999.

[3] A. Frank, S. Tanner, and P. Pevzner, "Peptide Sequence Tags for Fast Database Search in Mass Spectrometry," *International Conference on Research in Computational Molecular Biology (RECOMB)*, 2005.

[4] V. Dancik, T. Addona, K. Clauser, et al., "De novo protein sequencing via tandem mass-spectrometry," *Journal of Computational Biology*, vol. 6, pp. 327-341, 1999.

[5] A. Frank and P. Pevzner, "PepNovo: De Novo Peptide Sequencing via Probabilistic Network Modeling," *Analytical Chemistry*, vol. 77, pp. 964 -973, 2005.

[6] B. Ma, K. Zhang, C. Hendrie, et al., "PEAKS: Powerful Software for Peptide De Novo Sequencing by MS/MS," *Rapid Comm in Mass Spectrometry*, vol. 17, pp. 2337-2342, 2003.

[7] J. A. Taylor and R. S. Johnson, "Sequence database searches via de novo peptide sequencing by tandem mass spectrometry," *Rapid Comm in Mass Spec*, vol. 11, pp. 1067-1075, 1997.

[8] S. Tanner, H. Shu, A. Frank, et al., "InsPecT: Fast and accurate identification of post-translationally modified peptides from tandem mass spectra," *Analytical Chemistry*, vol. 77, pp. 4626-4639, 2005.

[9] K. Ning, H. K. Ng, and H. W. Leong, "PepSOM: An Algorithm for Peptide Identification by Tandem Mass Spectrometry based on SOM," *Genome Informatics*, vol. 17, pp. 194-205, 2006.

[10] H. K. Ng, K. Ning, and H. W. Leong, "A New Approach for Similarity Queries of Biological Sequences in Databases," *Pacific-Asia Conference on Knowledge Discovery and Data Mining (PAKDD)*, 2007.

[11] K. Ning, K. F. Chong, and H. W. Leong, "De novo Peptide Sequencing for Multi-charge Mass Spectra based on Strong Tags," *Asia Pacific Bioinformatics Conference*, 2007.

[12] T. Kohonen, *Self-Organizing Maps*, 3rd ed: Springer, 2001.

[13] H. K. Ng and H. W. Leong, "Path-Based Range Query Processing Using Sorted Path and Rectangle Intersection Approach," *International Conference on Database Systems for Advanced Applications (DASFAA)*, pp. 184-189, 2004.

[14] H. K. Ng, H. W. Leong, and N. L. Ho, "Efficient Algorithm for Path-Based Range Query in Spatial Databases.," *Int'l Database Engg & Applications Symp (IDEAS)*, pp. 334-343, 2004.

[15] D. Tsur, S. Tanner, E. Zandi, et al., "Identification of post-translational modifications by blind search of mass spectra," *Nature Biotechnology*, vol. 23, pp. 1562 - 1567, 2005.

[16] T. Kohonen, J. Hynninen, J. Kangas, et al., "SOM_PAK: The Self-Organizing Map Program Package," *Technical Report A31*, pp. FIN-02150 Espoo, 1996.

[17] J. T. Prince, M. W. Carlson, R. Wang, et al., "The need for a public proteomics repository," *Nature Biotechnology*, vol. 22, pp. 471-472, 2004.

[18] F. Desiere, E. W. Deutsch, N. L. King, et al., "The PeptideAtlas Project," *Nucleic Acids Research*, vol. 34, pp. D655-D658, 2006.

[19] A. Keller, S. Purvine, A. I. Nesvizhskii, et al., "Experimental protein mixture for validating tandem mass spectral analysis," *Omics*, vol. 6, pp. 207-212, 2002.

[20] S. Kim, S. Na, J. W. Sim, et al., "MODi: a powerful and convenient web server for identifying multiple post-translational peptide modifications from tandem mass spectra," *Nucleic Acids Research*, vol. 34, pp. W258-63, 2006.

[21] K. Kubota, T. Yoneyama-Takazawa, and K. Ichikawa, "Determination of sites citrullinated by peptidylarginine deiminase using 18O stable isotope labeling and mass spectrometry," *Rapid Communications in Mass Spectrometry*, vol. 19, pp. 683-8, 2005.

[22] S. R. Ramakrishnan, R. Mao, A. A. Nakorchevskiy, et al., "A fast coarse filtering method for peptide identification by mass spectrometry," *Bioinformatics*, vol. 22, pp. 1524-1531, 2006.

THE COMPARATIVE GENOMICS OF PROTEIN INTERACTIONS

JOSÉ M. PEREGRÍN-ALVAREZ[1,2]   CHRISTOS A. OUZOUNIS[3]
peregrin@ebi.ac.uk   ouzounis@ebi.ac.uk

[1]*Sick Kids Research Institute, TMDT MARS Building, 101 College St., 15th Floor, East Tower, M5G 1L7 Toronto, ON, Canada.*
[2]*Dpt. Of Molecular Biology and Biochemistry, University of Malaga, 29071 Malaga, Spain.*
[3]*Institute of Agrobiotechnology, Hellas 6th Km Charilaou, Thermi Rd, P.O. Box 361, 570 01 Thermi, Thessaloniki, Greece.*

The detection of gene fusion events across genomes can be used for the prediction of functional associations of proteins, including physical interactions or complex formation. These predictions are obtained by the detection of similarity for pairs of 'component' proteins to 'composite' proteins. Since the amount of composite proteins is limited in nature, we augment this set by creating artificial fusion proteins from experimentally determined protein interacting pairs. The goal is to study the extent of protein interaction partners with increasing phylogenetic distance, using an automated method. We have thus detected component pairs within seven entire genome sequences of similar size, using artificially generated composite proteins that have been shown to interact experimentally. Our results indicate that protein interactions are not conserved over large phylogenetic distances. In addition, we provide a set of predictions for functionally associated proteins across seven species using experimental information and demonstrate the applicability of fusion analysis for the comparative genomics of protein interactions.

Keywords: comparative genomics, protein interactions, *Escherichia coli*, *Helicobacter pylori*, two-hybrid screening.

1. Introduction

It has been shown that it is possible to predict protein interactions or, more generally, functional associations of proteins, including physical interaction or complex formation, using genome sequence analysis [1-3]. Fused genes encoding a single multifunctional protein in one species tend to be found in other species as pairs of genes encoding proteins showing similar functions or forming protein complexes [4]. Gene fusion is a well-known process in molecular evolution [5]; consequently, computational methods were developed to determine gene fusion events in complete genomes aiming to predict functional associations of proteins [1]. Many of these gene fusion events appear to be selectively advantageous by decreasing the regulational load in the cell for a particular process [1,3,5]. Thus, the detection of fused genes in one genome (defined as 'composite' proteins) allows the prediction of functional associations between homologous genes that remain separate in another genome (defined as 'component' proteins) [6].

Although gene fusion events (composite proteins) appear to be relatively rare [6], the accurate detection of a gene fusion event in one genome allows interactions to be predicted between many proteins across other genomes [1]. It is this kind of one-to-many relationship what makes this concept unique for discovering possible interactions or functional associations between proteins, even for those of unknown function, using comparative genomics. Unlike other methods that rely on gene proximity to predict functional coupling [7], the gene fusion method can also detect functional relationships of distal genes within a genome. Furthermore, we have previously demonstrated the high precision of the gene fusion method using the DIFFUSE algorithm [1] (see methods and Figure 1 for a flowchart of the algorithm), which with an additional constraint of minimum alignment overlap [6] has increased to over 86% (see Methods). This computational method is analogous and complementary to the experimental approaches for the detection of protein interactions [8].

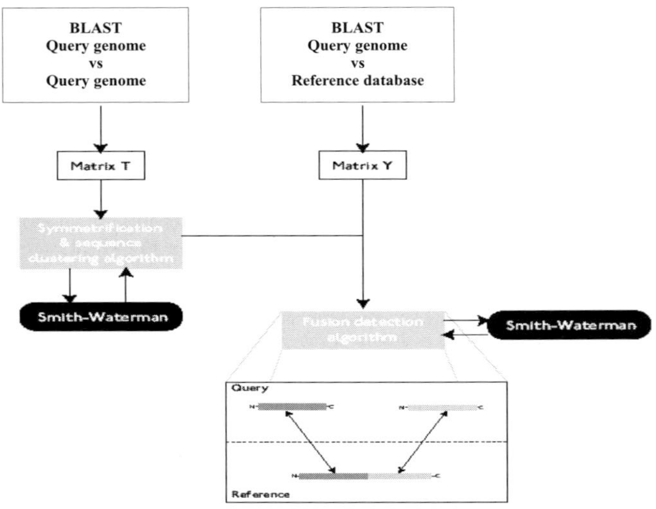

Figure 1. Flowchart of the Diffuse algorithm [1]. Similarities within the query genome using BLAST [12] are stored in a matrix T. An additional Smith-Waterman comparison [22] is used to resolve false negatives. The query genome is then compared to the reference genome, and similarities are stored in matrix Y. The fusion-detection algorithm identifies cases of the form depicted in the inset, where query proteins A and B exhibit similarity to reference protein C by checking matrix Y, but not to each other by checking matrix T, which is further confirmed by an additional Smith-Waterman comparison. Both Smith-Waterman runs are executed an additional 25 times, with randomization of the sequences, and a Z-score metric is obtained: if the Z-score is higher than certain threshold, the similarity is accepted as significant.

To examine the phylogenetic distribution (i.e conservation) of experimentally obtained protein interactions and generate predictions using comparative genomics, we have performed fusion analysis for seven entire genome sequences of similar size. Using DIFFUSE, we asked the question whether pairs of interacting proteins in *Helicobacter pylori* [9] are conserved across seven species of increasing phylogenetic distance (see methods). To achieve this, all pairs of interacting proteins from *H. pylori* have been merged to create a set of artificial fusion proteins. We define the genome where we seek component proteins as the 'query' genome and the set of artificially merged sequences from which we obtain artificial composite proteins as the 'reference' genome [1,6]. An 'artificial fusion event' is therefore defined as any pair of component proteins from a query genome that are detected as a fused, artificial composite protein in a reference genome (Figure 1).

DIFFUSE was applied individually for each of the seven genomes, against the artificially merged sequences which are used as the reference set (see Methods). Paralogy in the query genome makes it difficult to determine precisely the actual number of possible associations and increases uncertainty in accurately predicting component proteins [1]. However, the detection of component pairs, in different species with similarly sized genomes, via their similarity to interacting pairs in *H. pylori* followed by and additional constraint of minimal alignment overlap, allows us to assess the extent of the conservation of interacting proteins and generate predictions of protein interactions for distantly related species.

2. Methods

To generate predictions of functional associations in complete genomes, we have extracted the information and sequences of interacting proteins (6,797 sequences involved in 11,251 protein interactions) from the Database of Interacting Proteins (DIP) [10]. Of these interactions, 13% refer to *Helicobacter pylori* and were subsequently used in this analysis. We artificially merged sequence pairs using their original DIP binary relationships in order to get a Reference data set (Figure 1).

All selected genome sequences were obtained from their original sources [11] and used as queries. Genomes were selected based on phylogenetic distance to the reference genome *H. pylori*. Phylogenetic distance was defined by phylogenetic depth in a species-tree built by using Small Subunit rRNAs from the Ribosomal Database Project (RDP) [http://rdp.cme.msu.edu/html/].

The query database is compared against itself using BLASTp [12] (E-value threshold 10^{-06}), after masking compositional biased regions with the CAST [13] algorithm (score threshold 40), and all pairwise sequence similarities are recorded in a binary matrix. The query database is also compared against the reference database, as above, and similarities are recorded in another binary matrix. The DIFFUSE algorithm [1] (see Figure 1 for a flowchart of the algorithm) was then applied to both matrices and the detected 'artificial' gene fusion results were further filtered for significant overlap by more than 10% of their total length when aligned together with the artificial composite protein [6].

To assess the quality of the interaction and prediction data, we filtered all interacting pairs using either a functional class or subcellular localization criteria. We created 3 categories for functional classes (identical or positive, different, and unknown class), and 2 categories for subcellular localization (identical or positive, and different). The information about functional classes and subcellular localization was extracted from GeneQuiz [14] and MIPS [15], respectively.

The total automatic analysis was performed over a period of 72 hours on a 4-CPU Sun E450 with 2GB of RAM.

3. Results

As a first estimate of the performance of our approach and in order to assess the quality of our predictions, we have first tested a *H. pylori*-related species, the complete genome of the Proteobacteria *Escherichia coli* (4,290 ORFs) as query, against 1,359 *Helicobacter pylori* artificially fused sequences, defined as the reference set. We thus detected possible interacting partners using the DIFFUSE algorithm [1,6] and subsequently made predictions of functional associations of proteins across species. True positive protein interactions are expected to involve protein partners that belong to the same functional class [16]. Hence, we tested whether the artificial fusion events and component proteins identified by the DIFFUSE algorithm tend to involve component proteins with similar functional annotation. Thus, whenever a possible artificial fusion event is found, the artificial composites and the two components detected are assigned to 3 classes of functional information (see Methods). Conceptually, this approach is similar to the comparative analysis of protein interactions for *H. pylori* and *E. coli* [17]. Our analysis yields 1,487 pairs of *E. coli* proteins, with 141 (9%) classified as positives cases (i.e. in the same functional class)(see methods), 711 (48%) in different classes and 635 (43%) have at least one component with no functional class assignment (unknown) (see Methods).

To enhance the quality of our predictive analysis and eliminate noise from experimental procedures, we only consider components classified in the same functional class, thus all observations of different classes or those not fully classified are not further considered. This assumption obviously decreases the predictive potential of our method because it ignores cases of pairs of interacting proteins without class assignment, which may be functionally related [1,16]. The detected component proteins are far fewer in number than for the five previous reports of *E. coli* [1,3,6,16,17], due to the much more stricter criteria employed in this study and the multi-step protocol we have developed. Of these 141 positive cases only 12 appear to represent the same pairs of interacting proteins (putative orthologs)(see methods) in both the query (components) and reference set (artificial composites) (Table 1). Our method identifies a number of well-known interacting protein pairs. These are proteins participating in the same protein complex or biochemical process, such as Regulatory functions, Replication, Transcription, Translation and Transport-and-Binding proteins, according to the GeneQuiz functional classification (formed by 15 different functional classes) [14]. A number of unconfirmed cases constitute some interesting testable predictions. For example the phosphate regulon transcriptional regulatory protein PhoB was predicted to interact with the chemotaxis protein CheA (see URL in Discussion). The other 129 cases do not have consistent annotations across the two species because they are either paralogous genes or share specific domains with one or both of the artificial fused genes.

Coverage cannot easily be estimated, as we do not know in advance how many proteins potentially interact within the query genome. Thus, we have tried to estimate coverage by using the *H. pylori* genome as query against the *H. pylori* artificially fused sequences reference set, as described above, and then counting the number of predicted artificial composites.

Table 1. Prediction of protein interactions in *E. coli* using artificially fused *H. pilory* sequences.

S	FC	SI	ID	FUNCTION	SI	ID	FUNCTION
R	Rg	HP1067	CHEY_HELPY	Chemotaxis protein CheY	HP0392	O25153	Chemotaxis protein CHEA (EC 2.7.3.-)
P		1788191	CHEY_ECOLI		1788197	CHEA_ECOLI	
R	Rp	HP0705	UVRA_HELPY	Excinuclease ABC subunit A	HP1541	MFD_HELPY	Transcription-repair coupling factor (TRCF)
P		2367343	UVRA_ECOLI		1787357	MFD_ECOLI	
R	Rp	HP0705	UVRA_HELPY	Excinuclease ABC subunit A	HP1114	UVRB_HELPY	Excinuclease ABC subunit B
P		2367343	UVRA_ECOLI		1786996	UVRB_ECOLI	
R	Rp	HP0705	UVRA_HELPY	Excinuclease ABC subunit A	HP0821	UVRC_HELPY	Excinuclease ABC subunit C
P		2367343	UVRA_ECOLI		1788221	UVRC_ECOLI	
R	Tc	HP1293	RPOA_HELPY	RNA polymerase alpha subunit	HP1198	O25806	RNA polymerase beta chain (EC 2.7.7.6)
P		1789690	RPOA_ECOLI		1790419	RPOB_ECOLI	
R	Tl	HP0399	RS1_HELPY	30S ribosomal protein S1	HP1048	IF2_HELPY	Translation initiation factor IF-2
P		1787140	RS1_ECOLI		1789559	IF2_ECOLI	
R	Tl	HP1246	RS6_HELPY	30S ribosomal protein S6	HP1244	RS18_HELPY	30S ribosomal protein S18
P		1790644	RS6_ECOLI		1790646	RS18_ECOLI	
R	Tl	HP1246	RS6_HELPY	30S ribosomal protein S6	HP0886	SYC_HELPY	Cysteinyl-tRNA synthetase (EC 6.1.1.16)
P		1790644	RS6_ECOLI		1786737	SYC_ECOLI	
R	Tl	HP1312	RL16_HELPY	50S ribosomal protein L16	HP0083	RS9_HELPY	30S ribosomal protein S9 (BS10)
P		1789709	RL16_ECOLI		1789625	RS9_ECOLI	
R	Tl	HP1312	RL16_HELPY	50S ribosomal protein L16	HP1316	RL2_HELPY	50S ribosomal protein L2
P		1789709	RL16_ECOLI		1789713	RL2_ECOLI	
R	Tp	HP0687	O25396	Ferrous iron transport protein B	HP1072	COA0_HELPY	Copper-transporting ATPase (EC 3.6.1.36)
P		1789813	FEOB_ECOLI		1786691	ATCU_ECOLI	
R	Tp	HP0687	O25396	Ferrous iron transport protein B	HP1506	O26036	Sodium/Glutamate symport carrier protein
P		1789813	FEOB_ECOLI		1790085	GLTS_ECOLI	

The 12 positive pairs with identical functional class assignments (see Methods): pairs of interacting proteins in the query (*E. coli* components) and reference set (*H. pylori* artificial composites). Column names: Source (S) of information divided into Reference set (R) and the prediction for *E. coli* (P); Functional Class (FC) (Rg, Regulatory functions; Rp, Replication; Tc, Transcription; Tl, Translation; and Tp, Transport-and-binding proteins); Identifier (ID) from Swissprot and Fuctional assignment (Function), Sequence Identifier (SI), according to Genequiz [14]. Table is sorted by Functional class. Empty cells in Function columns, for simplicity, imply identical assignment between Reference and Prediction. Columns 3-5 and 6-8 correspond to the details of the individual component proteins. Predictions were performed with the DIFFUSE algorithm [1,6] using the *H. pylori* artificially merged sequences as Reference set and the *E. coli* genome as query (see Figure 1).

This number is equal to the number of composites in the original reference set, i .e. we obtained a 100% coverage when self-interactions are removed from the original DIP [10] data set, and 96% coverage when self-interactions are not removed, since DIFFUSE cannot detect sequence-similar components (see methods) [1,6]. We then calculated the percentage of interacting proteins (components) that are shared by the two species, defined as the number of unique (non-redundant) detected components in *E. coli* vs. the *H. pylori* reference set, divided by the number of unique components in a control experiment of the *H. pylori* genome as query vs. *H. pylori* reference set. This yielded a 60% of conserved protein pairs. This number strongly depends on the phylogenetic distance between query and reference genomes [6] but, a priori, it suggests that protein interactions (as pairs of proteins) may not be strongly conserved across related species.

To investigate the conservation of interacting pairs across other genomes in a more consistent way, we have repeated this analysis across other six species with similar genome sizes. Previously, we have explored the influence of three key factors in gene fusion analysis: genome size, paralogy and phylogenetic distance [6]. Herein, we focus on the latter, to understand the conservation of protein interactions from a comparative genomics perspective. To investigate further the results obtained from the comparison with *E. coli* (see above), we used an additional six genomes of similar size: *Campylobacter jejuni* (1,634 ORFs), *Haemophilus influenzae* (1,707 ORFs), *Borrelia burgdorferi* (1,639 ORFs), *Streptococcus pyogenes* (1,696 ORFs), *Thermotoga maritima* (1,849 ORFs) and *Thermoplasma acidophilum* (1,478 ORFs); plus the *H. pylori* genome (1,575 ORFs) as control. Since the number of components detected in each species depends on genome size, we have selected genomes according to two criteria: roughly similar genome size as the *H. pylori* genome (ranging between 1,478 and 1,849 ORFs) and a wide range of phylogenetic distances to *H. pylori* (Figure 2). As expected, there is an inverse relationship between the number of components and phylogenetic distance. This trend is only violated in the case of the *S. pyogenes* and *T. maritima* genomes, partly explained by the degree of paralogy for certain proteins [1]. The key conclusion from this analysis is that interactions as 'pairs' of protein partners seem are not highly conserved, are eroded over large phylogenetic distances, and may correspond to species-specific instances of interacting pairs (Figure 2).

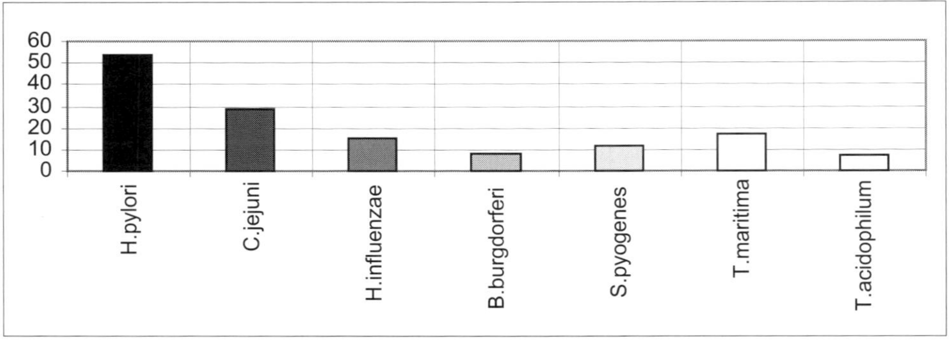

Figure 2. Numbers of protein interactions detected in seven genomes. Genomes are shown on the x-axis and the relative number of artificial component pairs (putative homolog interactions) on the y-axis. Genomes are sorted by decreased phylogenetic distance (see methods) to the *H. pylori* genome, which was used as control set. Bar shading from black to white represents phylogenetic distance to *H. pylori* (last two bars with white color for simplicity). Counts are normalized by genome size, although genomes sizes are comparable (actual counts are shown in Table 2A, row: TOTAL).

How many of the conserved pairs are predicted to belong to the positive class (i.e. same functional class)? When we perform the same analysis using functional class assignments, it is possible to associate those with phylogenetic distance (Table 2A). Again, generally speaking, the trend of decreasing number of interactions over large phylogenetic distances is observed, although less clearly – due to various degrees of annotation accuracy, obtained from GeneQuiz [14]. We again focus on positive cases (see methods) where predictions are of the highest quality (Table 2B). Among all predicted positive cases only predictions assigned to the Transcription class, namely between RNA polymerase subunits, are detected in 6 out of the 7 genomes. The method cannot predict the same pair of interacting proteins in the *Thermoplasma acidophilum* genome, due to the highly divergent nature of the archaeal RNA polymerase. This fact further illustrates the point that only a few interactions seem to be highly conserved or detectable across large evolutionary distances (see discussion).

Table 2. Prediction of protein interactions in complete genomes. Predictions were performed by the DIFFUSE algorithm (see Figure 1) [1] using the *H. pylori* artificially fused set as Reference and 6 different genomes as queries. **(A)** Number of components pairs in different genomes. The *H. pylori* genome was used as a control. Column names: Categories according to their distribution of functional classes (see Methods) followed by species names (*). Species (columns 2 to 8) are sorted by phylogenetic distance to *H. pylori*. **(B)** Potential positive examples of pairs of interacting proteins in the query genomes (components) and reference set (artificial composites) are listed.

(A)

Categories	*H. pylori*	*C. jejuni*	*H. influenzae*	*B. burgdorferi*	*S. pyogenes*	*T. maritima*	*T. acidophilum*
Same functional class	41	48	39	14	26	30	15
Different functional class	292	300	284	54	293	173	101
At least one unknown functonal class	4461	935	196	195	285	600	70
TOTAL	4794	1283	519	263	604	803	186

(B)

S	FC	SI	ID	FUNCTION	SI	ID	FUNCTION
R	Tc	HP1293	RPOA_HELPY	RNA polymerase alpha	HP1198	O25806	RNA polymerase beta
Hp	Tc	HP1293	RPOA_HELPY		HP1198	O25806	
Cj	Tc	6969012	RPOA_CAMJE		6967949	RPOB_CAMJE	
Hi	Tc	HI0802	RPOA_HAEIN		HI0515	RPOB_HAEIN	
Bb	Tc	BB0502	RPOA_BORBU		BB0389	RPOB_BORBU	
Sp	Tc	13621394	RPOA_STRPY		13621404	Q9A1U1	
Tm	Tc	TM1472	RPOA_THEMA		TM0458	RPOB_THEMA	
Cj	Tc	6969012	RPOA_CAMJE	RNA polymerase alpha	6967950	RPOC_CAMJE	RNA polymerase beta'
Hi	Tc	HI0802	RPOA_HAEIN		HI0514	RPOC_HAEIN	
Bb	Tc	BB0502	RPOA_BORBU		BB0388	RPOC_BORBU	
Sp	Tc	13621394	RPOA_STRPY		13621405	RPOC_STRPY	
Tm	Tc	TM1472	RPOA_THEMA		TM0459	RPOC_THEMA	
Cj	Tc	6967949	RPOB_CAMJE	RNA polymerase beta	6967950	RPOC_CAMJE	RNA polymerase beta'
Hi	Tc	HI0515	RPOB_HAEIN		HI0514	RPOC_HAEIN	
Bb	Tc	BB0389	RPOB_BORBU		BB0388	RPOC_BORBU	
Sp	Tc	13621404	Q9A1U1		13621405	RPOC_STRPY	
Tm	Tc	TM0458	RPOB_THEMA		TM0459	RPOC_THEMA	
Ta	Tc	10639562	RPOB_THEAC	RNA polymerase B	10639564	RPA2_THEAC	RNA polymerase A"

Column names: as in Table 1. Note that only the first case corresponds to the reference set, while the other cases are identified due to paralogy between the corresponding components. (*) Abbreviations: Hp, *Helicobacter pylori*; Cj, *Campylobacter jejuni*; Hi,*Haemophilus influenzae*; Bb, *Borrelia burgdorferi*; Sp, *Streptococcus pyogenes*; Tm, *Thermotoga maritima*; and Ta, *Thermoplasma acidophilum*.. Tc, Transcription class.

In order to assess the quality of the original data from DIP [10] in terms of predicting functional associations, we analyzed the patterns of distribution of Functional Classes for protein pairs (in terms of positive, different or unknown classes)(see methods) according to three different annotation schemes: Clusters of Orthologous Groups (COGs) [18], Euclid [19] and GeneQuiz [14]. The COG scheme yields more cases in the same functional class compared to the other two schemes in relative terms (9% of interactions in the reference set), although only 1106 sequences out of the total 1575 sequences from *H. pylori* genome are assigned to COGs. Therefore, we opted using the GeneQuiz functional class scheme (2%), since it provides maximum coverage of the genome, similarly to Euclid (3%) [19].

We then assessed the patterns of distribution per functional class in the positive category (see methods) for the six selected genomes and the reference set (Figure 3). In general, functional class distribution of positive pairs exhibits a highly non-uniform pattern for all six species examined and the reference set. The three more abundant functional classes correspond to the transport-and-binding-proteins, replication and translation, in this order, whose pattern is different from the reference set, suggesting that the observed interactions in *H. pylori* are not conserved in terms of functional class. The cell envelope functional class, which is one of the most abundant classes in the reference set, does not give rise to any predictions across the selected six genomes except for the case of the *H. pylori* genome used as control.

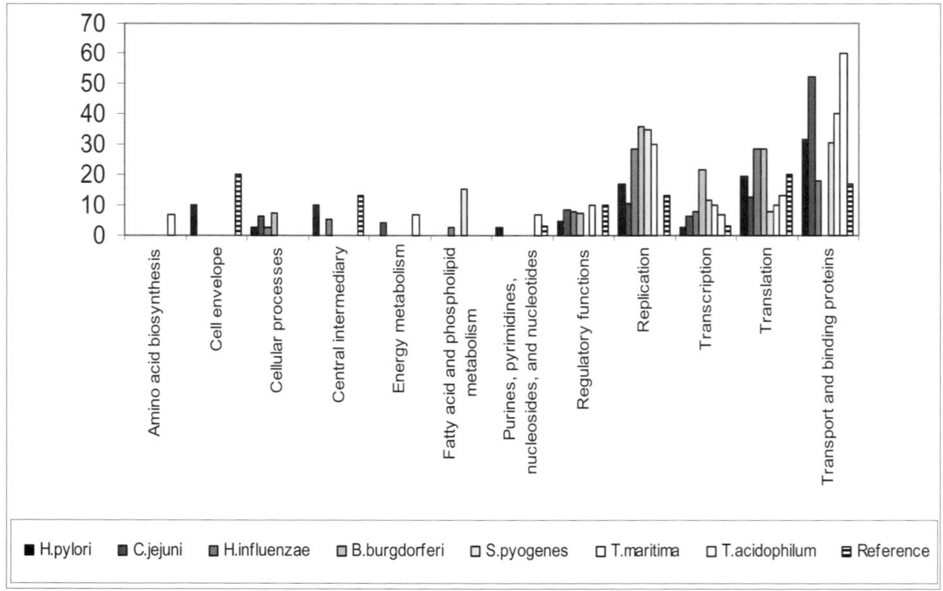

Figure 3. Relative distribution of functional classes for pairs of component proteins in the positive category. Functional classes are shown on the x-axis and the number of component pairs (as %) per functional class on the y-axis. Genomes are sorted by phylogenetic distance to the *H. pylori* genome, which was used as a control set. Functional class are sorted out by alphabetical order. Bar shading as in Figure 2 but the reference set which is shown as dark horizontal.

4. Discussion

Is there any bias in protein interaction information stemming from specific experimental techniques? Given the fact that all the *H. pylori* data comes exclusively from two-hybrid screening tests [9], we examine the influence of other experimental techniques for protein interaction detection. We counted the number of protein interactions per method (available in DIP), finding out that there are only two (out of 38) methods with more than 1,000 interactions per method (out of 11,251 recorded protein interactions in total). These two methods are two-hybrid screening and immunoprecipitation with 9,781 and 1,243 protein interactions, respectively. The remaining cases all have counts less than 400 interactions per method for a variety of species.

In fact, the only species (out of 112 species represented in DIP) for which there is sufficient information of protein interactions obtained by more than one method, including two-hybrid screening and immunoprecipitation is *Saccharomyces cerevisiae*. Protein interactions from this species can be validated not only by functional classes but also by subcellular localization. When all interacting proteins from *S. cerevisiae* are extracted and classified according to the corresponding experimental method, immunoprecipitation appears to yield more consistent functional classes (28% of interactions) and subcellular localization (75% of interactions), compared to two-hybrid screening (5% and 44%, respectively). Thus, there is a potential bias due to the experimental data used in our study. Furthermore, two-hybrid screening have been shown to have a high false positive (and false negative) rate [20]. Our results also show that subcellular localization is a highly desirable attribute for protein interactions that could be used both as a quality measure for the original DIP information as well as a filter for predicting reliable interactions. We should also note that only in a handful of cases, the two above mentioned experimental methods result in identical pairs of proteins, indicating that there is very little overlap of reliable experimental observations [21] (not shown).

Another caveat of our approach is that our results on the poor conservation of interactions across genomes may be due to difficulty of retrieving such relationships across large phylogenetic distances due to the use of BLAST [12] for protein similarity searches. Therefore, there is a possibility that the interaction pairs are indeed maintained across evolution but our approach failed to retrieve it. Furthermore, our approach rely on the quality and extent of the functional annotation data. Since our analysis shows that annotation schemes differ across genomes and it is biased towards identifying proteins of known function in more distantly related taxa (see Table 2) this might represent a bias in the extent and quality of the annotations used in this study.

In summary, although our approach has some caveats that will be addressed in a larger scale follow-up analysis, our preliminary analysis shows that the exhaustive detection of 'artificial' gene fusion events allows the prediction of functionally associated components based merely on genome structure. This approach for the prediction of functional associations of proteins results in accurate predictions for physical interactions, pathway involvement, complex formation and other types of functional associations of protein molecules, many of them may provide further support for previous studies [1,3,6,16,17].

All the cases presented here are based on the high-throughput protein interaction set from *H. pylori* and represent interesting and novel findings available at the following URL: http://cgg.ebi.ac.uk/old/cgg//projects/mining/artifuse/.

Acknowledgments

We thank members of the former Computational Genomics Group for discussions. This work was supported by the European Molecular Biology Laboratory and the Ministry of Science and Technology, Spain. C. A. O. thanks the UK Medical Research Council and IBM Research for additional support.

References

[1]   Enright, A.J., Iliopoulos, I., Kyrpides, N.C. and Ouzounis, C.A. (1999) Nature 402, 86-90.
[2]   Marcotte, E.M., Pellegrini, M., Thompson, M.J., Yeates, T.O. and Eisenberg, D. (1999) Nature 402, 83-6.
[3]   Marcotte, E.M., Pellegrini, M., Ng, H.L., Rice, D.W., Yeates, T.O. and Eisenberg, D. (1999) Science 285, 751-3.
[4]   Sali, A. (1999) Nature 402, 23, 25-6.
[5]   Doolittle, R.F. (1999) Nat Genet 23, 6-8.
[6]   Enright, A.J. and Ouzounis, C.A. (2001) Genome Biol 2(9):research0034.1-0034.7.
[7]   Overbeek, R., Fonstein, M., D'Souza, M., Pusch, G.D. and Maltsev, N. (1999) Proc Natl Acad Sci U S A 96, 2896-901.
[8]   Ito, T. et al. (2000) Proc Natl Acad Sci U S A 97, 1143-7.
[9]   Rain, J.C. et al. (2001) Nature 409, 211-5.
[10] Xenarios, I., Salwinski, L., Duan, X.J., Higney, P., Kim, S.M. and Eisenberg, D. (2002) Nucleic Acids Res 30, 303-5.
[11] Kyrpides, N.C. (1999) Bioinformatics 15, 773-4.
[12] Altschul, S.F., Madden, T.L., Schaffer, A.A., Zhang, J., Zhang, Z., Miller, W. and Lipman, D.J. (1997) Nucleic Acids Res 25, 3389-402.
[13] Promponas, V.J., Enright, A.J., Tsoka, S., Kreil, D.P., Leroy, C., Hamodrakas, S., Sander, C. and Ouzounis, C.A. (2000) Bioinformatics 16, 915-22.
[14] Andrade, M.A. et al. (1999) Bioinformatics 15, 391-412.
[15] Mewes, H.W. et al. (2002) Nucleic Acids Res 30, 31-4.
[16] Yanai, I., Derti, A. and DeLisi, C. (2001) Proc Natl Acad Sci U S A 98, 7940-5.
[17] Wojcik, J. and Schachter, V. (2001) Bioinformatics 17, S296-305.
[18] Tatusov, R.L. et al. (2001) Nucleic Acids Res 29, 22-8.
[19] Tamames, J., Ouzounis, C., Casari, G., Sander, C. and Valencia, A. (1998) Bioinformatics 14, 542-3.
[20] Deane C., Salwiński Ł., Xenarios I., Eisenberg D. (2002). Mol Cell Proteomics 1 (5): 349-56.

[21] von Mering, C., Krause, R., Snel, B., Cornell, M., Oliver, S.G., Fields, S. and Bork, P. (2002) Nature 417, 399-403.
[22] Smith T.F. & Waterman M. J. (1981). J Mol Biol 147, 195-197.

WEIGHTED LASSO IN GRAPHICAL GAUSSIAN MODELING FOR LARGE GENE NETWORK ESTIMATION BASED ON MICROARRAY DATA

TEPPEI SHIMAMURA
shima@ims.u-tokyo.ac.jp

SEIYA IMOTO
imoto@ims.u-tokyo.ac.jp

RUI YAMAGUCHI
ruiy@ims.u-tokyo.ac.jp

SATORU MIYANO
miyano@ims.u-tokyo.ac.jp

Human Genome Center, Institute of Medical Science, University of Tokyo, 4-6-1 Shirokanedai, Minato-ku, Tokyo 108-8639, Japan

We propose a statistical method based on graphical Gaussian models for estimating large gene networks from DNA microarray data. In estimating large gene networks, the number of genes is larger than the number of samples, we need to consider some restrictions for model building. We propose weighted lasso estimation for the graphical Gaussian models as a model of large gene networks. In the proposed method, the structural learning for gene networks is equivalent to the selection of the regularization parameters included in the weighted lasso estimation. We investigate this problem from a Bayes approach and derive an empirical Bayesian information criterion for choosing them. Unlike Bayesian network approach, our method can find the optimal network structure and does not require to use heuristic structural learning algorithm. We conduct Monte Carlo simulation to show the effectiveness of the proposed method. We also analyze *Arabidopsis thaliana* microarray data and estimate gene networks.

Keywords: Lasso; L_1-Regularization; Empirical Bayes; Graphical Modeling; Large Gene Network; DNA Microarray Data

1. Introduction

The interest in theoretical structure and constructive methodology for large-scale graphical models has been invigorated by problems of inferring gene networks from microarray gene expression data. Various reverse engineering methods for large gene networks have been proposed in the literature, and graphical Gaussian model has received a lot of attention recently as one of promising approaches for achieving the purpose [4, 14, 16, 19].

Several studies have used regression-based approaches to estimate structures of undirected graphs. The main idea is to apply regression analysis for each variable on the rest of the variables and the regression coefficients with relatively large absolute values correspond to edges in the graph. Meinshausen and Bühlmann [14] considered neighborhood selection as a subproblem of covariance selection and applied least absolute shrinkage and selection operator, called lasso [17], to estimate undirected graph structures and showed theoretical properties in high-dimensional situations, and Gustafsson *et al.* [6] also used the similar approach to estimate large gene networks from DNA microarray data.

However, recent works suggest that the lasso is not consistent for variable selection [11, 14, 21, 22]. That is, the sets of variables selected are not consistent at finding the true set of important variables. Meinshausen and Bühlmann [14] showed that the optimal regularization parameter for prediction accuracy does not lead to a consistent neighborhood estimate by giving an example in which the number of variables grows to infinity and in which the regularization parameter chosen according to prediction accuracy leads to the wrong model with probability tending to one. As the result, some noisy and irrelevant neighborhoods tend to be selected and the resulting graph is less sparse even if we use the lasso in the neighborhood selection problem.

We propose a new method for improving neighborhood selection in the graphical Gaussian model and define weighted lasso that allows different amount of penalty on each regression coefficient. We show that, by using the ridge estimator, the originally $(p-1)$ regularization parameter selection problem of each node j reduces to a univariate regularization parameter selection problem for λ_j. For choosing an optimal graph structure, we derived a new information criterion from an empirical Bayes viewpoint.

The rest of this paper is as follows. In Section 2, we describe our method. Especially, we review a class of graphical Gaussian graphical model and neighborhood selection for estimating sparse graph structure, and define a new neighborhood selection approach, in Section 2.1 and 2.2, respectively. In Section 2.3, we also derive a new criterion for choosing an optimal neighborhood for each node. In Section 3, we present the performance of the proposed method through numerical examples. In Section 3.1, We compare the performance of the proposed method with that of conventional graphical modeling approach through Monte Carlo simulation. We also apply our approach to *Arabidopsis thaliana* microarray data and estimate gene networks in Section 3.2. We provide concluding remarks in Section 4.

2. Method for Estimating Large Gene Networks

2.1. *Graphical Gaussian Model and Neighborhood Selection*

In this section, we present a class of graphical Gaussian model and review the neighborhood selection problem introduced by Meinshausen and Bühlman [14] that is a subproblem of covariance selection. A graphical Gaussian model or covariance selection model for the Gaussian random vector X is represented by an undirected graph $G = (V, E)$ with vertex set $V = \{1, \ldots, p\}$ and edge set $E = \{e^{ij}\}$, where $e^{ij} = 1$ or 0 according to whether vertices i and j are adjacent in G or not.

Let X be a p-dimensional random vector which consists of the p elements, $X^{(1)}, \ldots, X^{(p)}$. We assume that $X \sim N_p(\mathbf{0}, \mathbf{\Sigma})$ where $\mathbf{\Sigma}$ is a variance-covariance matrix. Our interest is to estimate the underlying graph of the distribution, that is to identify zero entries of the concentration matrix $C = \Sigma^{-1} = (c^{ij})$, because C satisfies the following linear restrictions:

$$e^{ij} = 0 \Rightarrow c^{ij} = 0. \tag{1}$$

The graphical Gaussian model also can be defined in terms of the pairwise conditional independence determined by the Markov properties of G. If $X \sim N_p(\mu, \Sigma)$, then

$$c^{ij} = 0 \Leftrightarrow X^{(i)} \perp\!\!\!\perp X^{(j)} \mid X^{V\setminus\{i,j\}} \Leftrightarrow \rho^{ij} = 0, \tag{2}$$

where

$$\rho^{ij} = \frac{-c^{ij}}{\sqrt{c^{ii}c^{jj}}}, \tag{3}$$

denotes the ij-th partial correlation, that is the correlation between the i-th variable and the j-th variable, say $X^{(i)}$ and $X^{(j)}$, conditional on the rest of the variables $X^{V\setminus\{i,j\}}$.

Given n random samples, $x_1 \ldots x_n$, of X where $x_i = (x_{i1}, \ldots, x_{ip})'$, the log-likelihood for $C = \Sigma^{-1}$ is

$$\frac{n}{2} \log |C| - \frac{1}{2} \sum_{i=1}^{n} (x_i - \mu)' C (x_i - \mu), \tag{4}$$

up to a constant not depending on μ and C. The unbiased estimator for the covariance matrix Σ is

$$\hat{\Sigma} = \frac{1}{n-1} \sum_{i=1}^{n} (x_i - \bar{x})(x_i - \bar{x})', \tag{5}$$

where $\bar{x}$ is the sample mean. Then, the concentration matrix C can be estimated by $\hat{\Sigma}^{-1}$. In general, the maximum likelihood estimator $\hat{C} = \hat{\Sigma}^{-1}$ contains no zero entry, and the network corresponding to $\hat{C}$ is thus the complete graph. However, the structure of the gene networks is known as a 'sparse' structure.

To achieve 'sparse' structure, edge selection is usually carried out by the following steps.

(1) Compute the sample concentration matrix $\hat{C}$ via $\hat{\Sigma}^{-1}$.
(2) Compute the sample estimates of partial correlation ρ^{ij} via equation (3).
(3) Determine which ρ^{ij} is 0 or not.

However, the sample covariance matrix $\hat{\Sigma}$ is singular in large gene network estimation from microarray data, and we cannot compute the partial correlation coefficients directly. To solve this problem, Meinshausen and Bühlmann [14] proposed the neighborhood selection method.

The neighborhood selection problem is introduced as a subproblem of covariance selection. Let the neighborhood ne_a of a node a be the smallest subset of $V \setminus \{a\}$ so that, given all variables X^{ne_a} in the neighborhood, $X^{(a)}$ is conditionally independent of all remaining variables. Denote the closure of node $a \in V$ by $\mathrm{cl}_a = \mathrm{ne}_a \cup \{a\}$. Then

$$X^{(a)} \perp\!\!\!\perp \{X^{(k)}; k \in V \setminus \mathrm{cl}_a\} | X^{\mathrm{ne}_a}. \tag{6}$$

Given $n \times p$ data matrix of X, $\boldsymbol{X} = (\boldsymbol{x}^{(1)}, \dots, \boldsymbol{x}^{(p)})$ where $\boldsymbol{x}^{(j)} = (x_1^{(j)}, \dots, x_n^{(j)})'$ is the n-dimensional sample vector of node j, the purpose of neighborhood selection is to estimate the neighborhood of any given node individually. We assume that the each column $\boldsymbol{x}^{(j)}$, $j = 1, \dots, p$ is centered and standardized such that $\sum_i x_i^{(j)}/n = 0$ and $\sum_i \{x_i^j\}^2/n = 1$. Meinshausen and Bühlmann [14] cast the neighborhood selection problem into a standard linear regression problem. The idea is to apply regression analysis for each variable on the rest of variables and the regression coefficients with nonzero values correspond to edges in the graph. They also show that this problem can be solved efficiently with the lasso [17]. For each node $j = 1, \dots, p$ on the rest of nodes, the lasso estimator $\hat{\boldsymbol{\theta}}^{[-j]} = (\hat{\theta}_j^{(1)}, \dots, \hat{\theta}_j^{(j-1)}, \hat{\theta}_j^{(j+1)}, \dots, \hat{\theta}_j^{(p)})'$ is given by

$$\hat{\boldsymbol{\theta}}^{[-j]} = \arg\min_{\theta^{[-j]}} \left\{ \frac{1}{n} \|\boldsymbol{x}^{(j)} - \boldsymbol{X}^{[-j]}\boldsymbol{\theta}^{[-j]}\|^2 + \lambda \sum_{k \neq j} |\theta_j^{(k)}| \right\}, \tag{7}$$

where $\boldsymbol{X}^{[-j]}$ is the $n \times (p-1)$ matrix resulting from the deletion of the j-th column from the data matrix $\boldsymbol{X}$ and λ is a non-negative regularization parameter. Then, the neighborhood estimate of node j is defined by

$$\hat{\mathrm{ne}}_j = \{k \in V; \hat{\theta}_j^{(k)} \neq 0\}. \tag{8}$$

Note that it is possible that $\theta_j^{(k)} = 0$ but $\theta_k^{(j)} \neq 0$, or $\theta_k^{(j)} = 0$ but $\theta_j^{(k)} \neq 0$. They suggest that the nodes j and k are connected by the undirected edge if $\hat{\theta}_j^{(k)} \neq 0$ or $\hat{\theta}_k^{(j)} \neq 0$.

2.2. *Improving Neighborhood Selection with Weighted Lasso*

In this section, we show a solution to overcome the drawback in the neighborhood selection of the graphical Gaussian model with the lasso. The major reason for the inconsistency of variable selection with the lasso comes from the fact that the lasso imposes the same amount of penalty on each regression coefficient, regardless of their relative significance [5, 22]. As a simple approach for allowing different amount of penalty on each regression coefficient, we consider the following problem for node j, $j = 1, \dots, p$:

$$\hat{\boldsymbol{\theta}}^{[-j]} = \arg\min_{\theta^{[-j]}} \left\{ \frac{1}{n} \|\boldsymbol{x}^{(j)} - \boldsymbol{X}^{[-j]}\boldsymbol{\theta}^{[-j]}\|^2 + \sum_{j \neq k} \lambda_j^{(k)} |\theta_j^{(k)}| \right\}. \tag{9}$$

Here $\lambda_j^{(k)}$, $k = 1, \dots, (j-1), (j+1), \dots, p$, are $(p-1)$ non-negative regularization parameters. Obviously, this problem depends on $(p-1)$ regularization parameters.

In practice, the choices of these regularization parameters are time-consuming, since we have to choose $(p-1)$ regularization parameters for each node j, which leads to the $p \times (p-1)$ regularization parameter selection problem for all nodes. To solve this problem, we propose to replace $\lambda_j^{(k)}$ by

$$\lambda_j^{(k)} = \frac{\lambda_j}{(\tilde{\theta}_j^{(k)})^{\tau}}, \tag{10}$$

where τ is a constant and $\tilde{\theta}_j^{(k)}$ is a weight for the significance of the k-th regression coefficient $\theta_j^{(k)}$. Let a $(p-1)$-dimensional weight vector by $\tilde{\boldsymbol{\theta}}_j = (\tilde{\theta}_j^{(1)}, \ldots, \tilde{\theta}_j^{(j-1)}, \tilde{\theta}_j^{(j+1)}, \ldots, \tilde{\theta}_j^{(p)})'$. We estimate $\tilde{\boldsymbol{\theta}}_j$ by

$$\tilde{\boldsymbol{\theta}}_j = (\boldsymbol{X}^{[-j]'}\boldsymbol{X}^{[-j]} + \gamma\boldsymbol{I}_{p-1})^{-1}\boldsymbol{X}^{[-j]}\boldsymbol{x}^{(j)}, \tag{11}$$

where γ is a pre-specified regularization parameter, which is equivalent to the ridge estimate. Even if the data matrix $\boldsymbol{X}^{[-j]}$ is rank-deficient, so that $\boldsymbol{X}^{[-j]'}\boldsymbol{X}^{[-j]}$ is singular, the regularized matrix $(\boldsymbol{X}^{[-j]'}\boldsymbol{X}^{[-j]} + \gamma\boldsymbol{I}_{p-1})$ is not singular for any nonzero value of γ. The regularization parameter γ also controls the number of neighborhoods. Although γ can be selected by some criteria such as cross-validation or our proposed criterion discussed in next section, we set γ a large value such as 10^{10}. Then, neighborhood selection proceeds by taking a node j to be a neighbor of node k if and only if $\hat{\theta}_j^{(k)} \neq 0$. In inferring network structure, the nodes j and k are connected by the undirected edge if $\theta_j^{(k)} \neq 0$ or $\theta_k^{(j)} \neq 0$. We call the successive procedure *weighted lasso*.

2.3. *A New Criterion for Regularization Parameter Selection*

The regularization parameters $\lambda_1, \ldots, \lambda_p$ for all nodes in (10) should be chosen in a reasonable manner. We investigate this problem from a statistical model evaluation point of view, and derive a new information criterion for neighborhood selection for each node j. We can choose the optimal neighborhoods for each node by using the derived information criterion.

It is known that the most of regularization approach have Bayesian interpretation, such that the loss function is interpreted as the negative log-likelihood, the penalty term as the negative log-prior density, the regularization parameter as the hyperparameter, and the regularized estimate corresponds to the maximum posteriori estimate. For node j, the loss function can be interpreted as the negative log-likelihood of the following Gaussian distribution

$$f(\boldsymbol{x}^{(j)}|\boldsymbol{x}^{[-j]}; \boldsymbol{\theta}^{[-j]}, \sigma_j^2) = \left(\frac{1}{2\pi\sigma_j^2}\right)^{n/2} \exp\left\{-\frac{1}{2\sigma_j^2}\|\boldsymbol{x}^{(j)} - \boldsymbol{X}^{[-j]}\boldsymbol{\theta}^{[-j]}\|^2\right\}, \tag{12}$$

and the prior density as the independent Laplace prior as

$$\pi(\boldsymbol{\theta}^{[-j]}|\boldsymbol{\lambda}^{[-j]}, \sigma_j^2) = \prod_{k \neq j} \pi(\theta_j^{(k)}|\lambda_j^{(k)}, \sigma_j^2) = \prod_{k \neq j} \left(\frac{\lambda_j^{(k)}}{2\sigma_j^2}\right) \exp\left\{-\frac{\lambda_j^{(k)}|\theta_j^{(k)}|}{\sigma_j^2}\right\}. \tag{13}$$

From an empirical Bayesian point of view, one can choose $\lambda_j^{(k)}$ by maximizing the marginal likelihood [1, 8, 9]. For node j, the marginal likelihood is computed by integrating over the unknown parameter values $\boldsymbol{\theta}^{[-j]}$ and is defined by

$$\mathrm{ML} = \int f(\boldsymbol{x}^{(j)}|\boldsymbol{x}^{[-j]}; \boldsymbol{\theta}^{[-j]}, \sigma_j^2)\pi(\boldsymbol{\theta}^{[-j]}|\boldsymbol{\lambda}^{[-j]}, \sigma_j^2)d\boldsymbol{\theta}^{[-j]}. \tag{14}$$

The marginal likelihood often contains a complicated integral for the parameters, which can be usually approximated by some approximation methods, for example, Laplace's method [18]. However, in situations where the some components of $\hat{\boldsymbol{\theta}}^{[-j]}$ are exactly to zero with L_1-regularization approaches such as the lasso, the functional in the integral (14) is not differentiable at the origin and thus the above approximation cannot be applied directly.

Let $A_j = \{k; \hat{\theta}_j^{(k)} \neq 0\}$ be *active set* of $\hat{\boldsymbol{\theta}}^{[-j]}$. To overcome such a problem, we define the following partial marginal likelihood given active set A_j by

$$\mathrm{PML} = \int f(\boldsymbol{x}^{(j)}|\boldsymbol{x}^{[-j]}; \boldsymbol{\theta}^{[-j]}, \sigma_j^2)\pi(\boldsymbol{\theta}^{[-j]}|\boldsymbol{\lambda}^{[-j]}, \sigma_j^2)d\boldsymbol{\theta}^{A_j}. \tag{15}$$

This quantity is computed by integrating over the unknown parameters $\boldsymbol{\theta}^{A_j}$ included with the set A_j. Suppose that $\mathrm{PML} = O(n)$ and λ depends on the number of sample n. Applying the Laplace method, we have

$$\mathrm{PML} \approx \left(\frac{2\pi}{n}\right)^{|A_j|/2} |\boldsymbol{H}|^{-1/2} \times f(\boldsymbol{x}^{(j)}|\boldsymbol{x}^{[-j]}; \hat{\boldsymbol{\theta}}^{A_j}, \sigma_j^2) \times \pi(\hat{\boldsymbol{\theta}}^{A_j}|\boldsymbol{\lambda}^{[-j]}, \sigma_j^2), \tag{16}$$

with $O(n^{-1})$ where $\hat{\boldsymbol{\theta}}^{A_j} = (\hat{\theta}_j^{(k)})_{k \in A_j}$ and $\boldsymbol{H}$ is a Hessian matrix given by

$$\boldsymbol{H} = -\frac{1}{n}\frac{\partial^2}{\partial\boldsymbol{\theta}^{A_j}\partial\boldsymbol{\theta}^{A_j}{}'}\left[\log\left\{f(\boldsymbol{x}^{(j)}|\boldsymbol{X}^{[-j]}; \boldsymbol{\theta}^{[-j]}, \sigma_j^2) \times \pi(\boldsymbol{\theta}^{[-j]}|\boldsymbol{\lambda}^{[-j]}, \sigma_j^2)\right\}\right]_{\boldsymbol{\theta}^{[-j]}=\hat{\boldsymbol{\theta}}^{[-j]}}.$$

Denote the right part of equation (16) by APML. Then, the estimator of $\hat{\sigma}_j^2$ is given by the equation

$$\frac{\partial}{\partial\sigma_j^2}\log(\mathrm{APML}) = 0. \tag{17}$$

Approximating $-2\log\mathrm{PML}$ and substituting $\hat{\sigma}_j^2$ for σ_j^2, we define an empirical Bayes criterion for neighborhood selection, called neighborhood empirical Bayes criterion (NEBC), as

$$\begin{aligned}
\mathrm{NEBC} = {}& (n + |A_j|)\left\{\log(2\hat{\sigma}^2) + 1\right\} - (n - |A_j|)\log\pi \\
& -2\sum_{k\in A_j}\log\lambda_j^{(k)} + \log|\boldsymbol{X}^{A_j}{}'\boldsymbol{X}^{A_j}|,
\end{aligned} \tag{18}$$

where

$$\hat{\sigma}_j^2 = \frac{\|\boldsymbol{x}_j - \boldsymbol{X}^{[-j]}\hat{\boldsymbol{\theta}}^{[-j]}\|^2 + 2\lambda_j\sum_{k\in A_j}|\hat{\theta}_j^{(k)}|}{n + |A_j|}. \tag{19}$$

For neighborhood selection problem for node j, we can choose λ_j by minimizing NEBC in (18) and estimate neighborhood $\hat{\mathrm{ne}}_j$ with $\hat{\boldsymbol{\theta}}^{[-j]}$.

3. Numerical Examples

3.1. *Monte Carlo Simulation*

We perform simulations to evaluate the performance of the proposed method and compare our method with the graphical Gaussian model approach with the lasso by Meinshausen and Bühlmann [14]. Considering finds on the topology of metabolic and protein networks [7, 13], we simulate a scale-free-like graphical model and generate data based on the model by the following steps.

(1) Generate a set of 20 scale-free networks by Barabási and Albert [3]. For each scale-free network, we start with a single node and no edges in the first time step. In each time step, one node is added and the new vertex initiates some edges to old nodes. The probability of node l, say Prob_l, that an old node is chosen is given by $\text{Prob}_l \propto k_l^\alpha$ where k_l is the in-degree of node l in the current time step, *i.e.*, the number of adjacent edges of node l which were not initiated by l itself. We set $k_l = 1$ and $\alpha = 1$. The node size of each scale-free network is set to 50. Then the total-node size of the simulated network is $20 \times 50 = 1000$.

(2) Generate a random variable from the resulting scale-free networks. As an initial value of the parent nodes of the 20 scale free-networks, we sample from $N_{20}(\mathbf{0}, \Sigma)$ where $\Sigma_{ii} = 1$ and $\Sigma_{i(i+1)} = \Sigma_{(i+1)i} = 0.5$ if $\{i; i = 1, \ldots, 19 \cap i \neq 4, 8, 12, 16\}$. From the j-th scale free network, we generate a random sample x based on

$$x \sim N(\sum_l w_l p_l(x), \sigma^2), \tag{20}$$

where $p_l(x)$ is the observation of the l-th parent of x and w_l is the coefficient. We sample w_l from the uniform distribution over the interval $[-1, 1]$ and set σ so that the signal-to-noise ratio is 0.1.

We simulate 100 observations with 1000 variables from the above process. The true graph and the estimated graph by the proposed method through the simulation are shown in Figure 1.

The performance of the graphical modeling approaches is evaluated by counting true positives (TP; correctly identified true edges), false positives (FP; spurious edges, that is not recognized as zero-edges), true negatives (TN; correctly identified zero-edges) and false negatives (FN; spurious zero-edges, that is not recognized as true edges). We perform 100 Monte Carlo simulations and calculate the means and standard deviations of TP, FP, TN and FN over these simulations. The result of comparison with the proposed method and the lasso-based neighborhood selection approach by Meinshausen and Bühlmann [14] is shown in Table 1. It can be seen from Table 1 that the neighborhood selection approach with the lasso [14] produces the larger number of false positives and identifies many unrelated edges as important edges. The proposed method reduces the false positives dramatically with a reasonable degree of true positives and succeeds in achieving the high proportion of true-edges within the small number of selected edges.

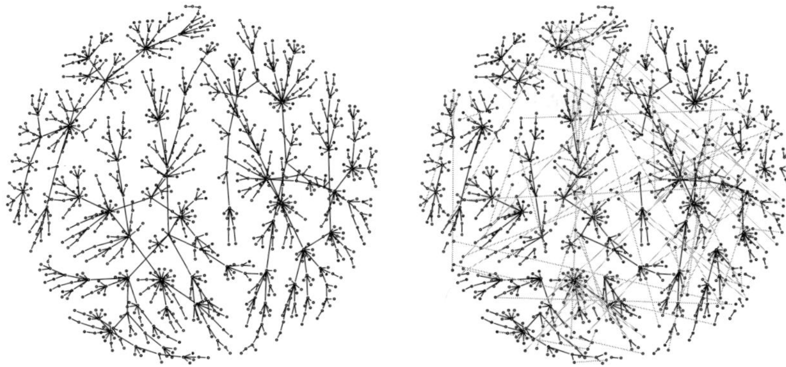

Fig. 1.   The left figure is the true graph generated for the Monte Carlo simulations. This network consists of 1000 nodes and 995 edges. The right figure is an example of the estimated network by the proposed method. The black lines describe the true-positive edges and the gray lines the false-positive edges.

Table 1.   Comparison of the averages and standard deviations of the true positives (TP), false negatives (FN), true negatives (TN) and false positives (FP) with the lasso-based neighborhood selection [14] and the proposed approach over the 100 simulations. The standard deviations are in parentheses.

Method	TP	FN	TN	FP
Lasso-based Method [14]	909.47 (6.01)	85.53 (6.01)	966630.22 (459.04)	32374.78 (459.04)
Proposed Approach	818.74 (6.29)	176.26 (6.29)	998894.65 (5.87)	110.35 (5.89)

3.2. *Example from Experimental Data*

We analyze the isoprenoid biosynthetic pathway data in *Arabidopsis thaliana* discussed by Wille *et al.* [20]. Wille *et al.* [20] reported a data set including the gene expression patterns monitored under various experimental conditions using 118 GeneChip microarrays.

It is known that plants contain two pathways, the mevalonate (MVA) pathway and the methylerythritol 4-phosphate (MEP) pathway, for the synthesis of the structural precursors of isoprenoids. To gain insights into the cross-talk between the MVA and MEP pathways at the transcriptional level and construct the gene network, Wille *et al.* [20] focuses on 39 genes where 15 genes were assigned to the cytosolic MVA pathway, 19 to the plastidal MEP pathway, and 5 genes encoding proteins located in the mitochondria.

We first estimate the undirected graph based on the 118 observations of these 39 gene expression profiles by the weighted lasso. Using NEBC in (18) for neighborhood selection, the proposed approach selects 26 gene-gene interactions. These 26 pairs are shown on the true network in Figure 2. We find there is a module with strongly connected genes in each of MVA pathway and MEP pathway. In the MVA pathway,

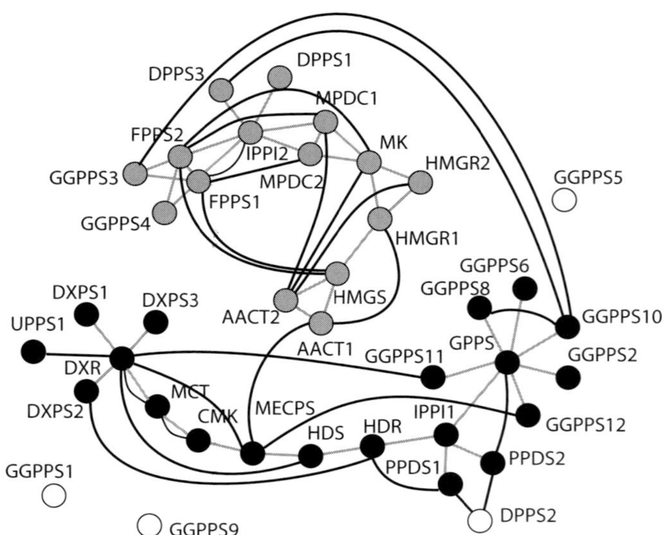

Fig. 2.   Pathways identified by the proposed method for the 39 genes in the isoprenoid pathway, where the grey color edges are the true pathways, and black curved edges are the estimated edges by the proposed method. The black nodes also represent a subgraph of the gene module in the MEP pathway, the gray nodes in the MVA pathway, and the white nodes in the Mitochondrion.

DXR, MCT, CMK and MECPS are connected as the known isoprenoid pathway. In the MEP pathway, AACT2, HMGR2, MK, MPDC1, FPPS1 and FPP2 are closely connected. Furthermore, several genes in the MEP pathway are linked to proteins in the mitochondria.

Wille *et al.* [20] also incorporated 795 additional gene expressions from 56 metabolic pathways and investigated which pathways attach significantly well to the MVA and MEP pathways. Among these were genes from pathways downstream of the two isoprenoid biosynthesis pathways, such as phytosterol biosynthesis, mono- and diterpene metabolism, porphyrin/chlorophyll metabolism, carotenoid biosynthesis, plastoquinone biosynthesis, for example. We applied our approach to the 118 observations of these total 835 gene from the 56 metabolic pathways, Mitochondrion, and the MVA and MEP pathways. Then we count the number of pathway-pathway interactions from the estimated graph and generate a "metabolic-pathway relevant network" shown in Figure 4. In the metabolic-pathway relevant network, nodes represent pathways, and two pathways are connected to each other if and only if they share at least 30 interactions.

We find that there are strong connections between the MEP pathway and the 8 other metabolic pathways; calvin cycle, carotenoid, citrate cycle, fatty acid, glycolysis, inositol phosphate, porphyrin chlorophyll, and sucrose pathways. On the other hand, the fatty acid, glycolysis, phytosterol, and sucrose pathways appear to be closely related to the genes of the MVA pathway. By comparing with the results

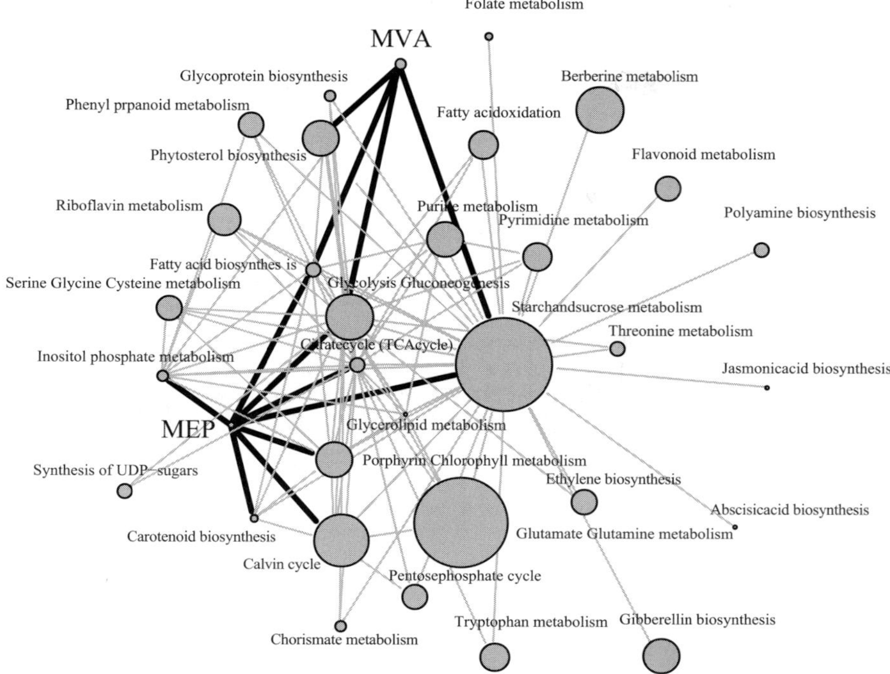

Fig. 3. The metabolic-pathway relevant network identified by the proposed method on 118 microarray data with 835 genes. The network contains the 56 metabolic pathways, mitochondrion, and the MVA and MEP pathways. Nodes represent pathways, and two pathways are connected to each other if and only if they share at least 30 interactions. The diameter of a circle is proportional to the number of genes participating in the corresponding pathways. The black edges represent the edges associated with the MVA and MEP pathways.

in Wille *et al.* [20], some of them are experimentally supported by [2, 10, 12, 15]. It is interesting that our method identifies the fatty acid pathway as a significant one related with both of the MVA and MEP pathways which was not recognized by Wille *et al.* [20].

4. Conclusion

We proposed the new method for improving neighborhood selection in the graphical Gaussian model by the weighted lasso that allows different amount of penalty on each regression coefficient. We showed that, by using the ridge estimator, the originally $(p-1)$ regularization parameter selection problem of each node j reduces to a univariate regularization parameter selection problem for λ_j. For choosing an optimal graph structure, we derived a new criterion from an empirical Bayes viewpoint.

We conducted Monte Carlo simulation and showed examples from microarray data in *Arabidopsis thaliana* to compare the proposed method with the other graphical Gaussian modeling approach with the lasso [14]. We found that our proposed method was superior to it.

References

[1] Akaike, H., Likelihood and the Bayes procedure, in J. M. Bernardo, M. H. DeGroot, D. V. Lindley and A. F. M. Smith (eds.), *Bayesian Statist.*, 143–166, University Press, Valencia, Spain, 1980.

[2] Arigoni, D., Sagner, S., Latzel, C., Eisenreich, W., Bacher, A. and Zenk, MH., Terpenoid biosynthesis from I-deoxy-D-xylulose in higher plants by intramolecular skeletal rearrangement, *Proc. Natl. Acad. Sci. USA*, 94:10600–10605, 1997.

[3] Barabási, A. and Albert, R., Emergence of scaling in random networks, *Science*, 286:509–512, 1999.

[4] Dobra, A., Jones, B., Hans, C., Nevis, J. and West, M., Sparse graphical models for exploring gene expression data, *J. Multivaliate Analysis*, 90:196–212, 2004.

[5] Fan, J. and Li, R., Variable selection via nonconcave penalized likelihood and its oracle properties, *J. Am. Statist. Assoc.* 96:1348–1360, 2001.

[6] Gustafsson, M., Hornquist, M. and Lombardi, A., Constructing and analyzing a large-scale gene-to-gene regulatory network-lasso-constrained inference and biological validation, *IEEE Trans. on Computational Biology and Bioinformatics*, 2:254–261, 2005.

[7] Jeong, H., Tombor, B., Albert, R., Oltvai, Z. and Barábasi, A., The large-scale organization of metabolic networks, *Nature*, 407:651–654, 2000.

[8] Kass, R. E. and Raftery, A. E., Bayes factor and model uncertainty, *J. Am. Statist. Assoc.*, 90:773–795, 1995.

[9] Konishi, S., Ando, T. and Imoto, S., Bayesian information criteria and smoothing parameter selection in radial basis function networks, *Biometrika*, 91:27–43, 2004.

[10] Laule, O., Fürholz, A., Chang, HS., Zhu, T., Wang, X., Heifetz, PB., Gruissem W. and Lange, M., Crosstalk between cytosolic and plastidial pathways of isoprenoid biosynthesis in *Arabidopsis thaliana*, *Proc. Natl. Acad. Sci. USA*, 100:6866–6871, 2003.

[11] Leng, C., Lin, Y. and Wahba, G., A note on the lasso and related procedures in model selection, *Statistica Sinica*, 16:1273–1284, 2006.

[12] Lichtenthaler, HK., Schwender, J., Disch, A. and Rohmer, M., Biosynthesis of isoprenoids in higher plant chloroplasts proceeds via a mevalonate-independent pathway, *FEBS Lett.*, 400:271–274, 1997.

[13] Maslov, S. and Sneppen, K., Specificity and stability in topology of protein networks, *Science*, 296:910–913, 2002.

[14] Meinshausen, N. and Bühlmann, P., High dimensional graphs and variable selection with the lasso, *Ann. Statist.*, 34:1436–1462, 2006.

[15] Rodriguez-Concepcion, M. and Boronat, A., Elucidation of the methylerythritol phosphate pathway for isoprenoid biosynthesis in bacteria and plastids. A metabolic milestone achieved through genomics, *Plant. Physiol.*, 130:1079–1089, 2003.

[16] Schäfer, J., Strimmer, K., An empirical Bayes approach to inferring large-scale gene association networks, *Bioinformatics*, 21:754–764, 2005.

[17] Tibshirani, R., Regression shrinkage and selection via the lasso, *J. R. Statist. Soc. B*, 58:267–288, 1996.

[18] Tierney, L. and Kadane, J. B., Accurate approximations for posterior moments and marginal densities, *J. Am. Statist. Assoc.*, 81:82–86, 1986.

[19] Toh, H. and Horimoto, K., Inference of a genetic network by a combined approach of

cluster analysis and graphical Gaussian modeling, *Bioinformatics* 18:287–297, 2002.

[20] Wille, A., Zimmermann, P., Vranová, E., Fürholz, A., Laule, O., Bleuler, S., Hennig, L., Prelić, A., Rohr, P. V., Thiele, L., Zitzler, E., Gruissem, W. and Bühlmann, Sparse graphical Gaussian modeling of the isoprenoid gene network in *Arabidopsis thaliana*, *Genome Biology*, 5:1-13, 2004.

[21] Yuan, M. and Lin, Y., On the nonnegative garrote estimator, *J. R. Statist. Soc.*, B, 69, 143–161, 2007.

[22] Zou, H., Adaptive lasso and its oracle properties, *J. Am. Statist. Assoc.*, 101:1418–1429, 2006.

GO based Tissue Specific Functions of Mouse using Countable Gene Expression Profiles

YOICHI TAKENAKA[1]   AKIKO MATSUMOTO[2]   HIDEO MATSUDA[1]

takenaka@ist.osaka-u.ac.jp   matsuda@ist.osaka-u.ac.jp

[1] *Department of Bioinformatic Engineering, Graduate School of Information Science and Technology, Osaka University Machikaneyama 1-3, Toyonaka City, Osaka 560-8531, Japan*
[2] *JASTEC Co., Ltd.,Takanawa 3-5-23, Minato-ku, Tokyo Japan*

We present a new method to describe tissue-specific function that leverages the advantage of the Cap Analysis of Gene Expression (CAGE) data. The CAGE expression data represent the number of mRNAs of each gene in a sample. The feature enables us to compare or add the expression amount of genes in the sample. As usual methods compared the gene expression values among tissues for each gene respectively and ruled out to compare them among genes, they have not exploited the feature to reveal tissue specificity. To utilize the feature, we used Gene Ontology terms (GO-terms) as unit to sum up the expression values and described specificities of tissues by them. We regard GO-terms as events that occur in the tissue according to probabilities that are defined by means of the CAGE. Our method is applied to mouse CAGE data on 22 tissues. Among them, we show the results of molecular functions and cellular components on liver. We also show the most expressed genes in liver to compare with our method. The results agree well with well-known specific functions such as amino acid metabolisms of liver. Moreover, the difference of inter-cellular junction among liver, lung, heart, muscle and prostate gland are apparently observed. The results of our method provide researchers a clue to the further research of the tissue roles and the deeper functions of the tissue-specific genes. All the results and supplementary materials are available via our web site.

Keywords: Tissue Specificity; Gene Ontology; Expression Profile; CAGE

1. Background

Analysis of gene expression has contributed to reveal the roles and functions of genes. Many researchers have studied genes with similar expression pattern [1] and have inferred the regulation networks of genes [2, 3]. The gene expressions are also used to reveal the roles and functions of tissues from the view point of genes. It is reasonable to infer that over-expressed genes in a tissue closely related to the functions of the tissue, and the genes that are not highly but expressed only in a particular tissue are also be related to. Hereafter, we call these tissue-specific genes. To observe gene expression, DNA microarrays [1, 4], EST (Expression Sequence Tag) [5], SAGE (Serial Analysis of Gene Expression) [6, 7], MPSS (Massively Parallel Signature Sequencing) [8], and CAGE (Cap Analysis of Gene Expression) [9, 10] has been widely used. Using these experimental data, Schug et al. found tissue-specific promoter features as measured by Shannon entropy [11], and Kadota et al. proposed a method to identify tissue-specific genes by outlier detection [12, 13]. When genes turn out to be tissue-specific, the functions of the genes help us to reveal the tissue-specific functions.

However, functions of the tissue-specific genes are insufficient to describe tissue-specific functions. Firstly, the genes that express significantly in a tissue not necessarily mean that the genes express higher than other genes. Of course, the tissue-specific genes with lower expression amount will be influenced to the inherent function of the tissue, but they can be lightly affected to than tissue-specific genes with higher expression amount. Secondly, it is natural that a set of genes contribute a role of tissues. For examples, hemoglobin is a hetero-dimer protein one of whose molecular function is oxygen binding. Biological process like chitin metabolisms and cellular components such as desmosome or connexson complex are closely-related to the functions of the tissue. As usual methods only focused on functions of genes, they wrote off the functions composed by a set of the genes. The aim of our study is to propose a method to determine, as completely as possible, tissue-specific functions that are hard to be identified from the tissue-specific genes described by usual methods.

To describe tissue-specific functions, we use Gene Ontology (GO) [14]. GO provides a set of structured vocabularies for specific biological domains that can be used to describe the domain in terms of their associated biological process, cellular components and molecular functions, in a species-independent manner. Each word is called a GO-term, and relationships among GO-terms are described as a directed acyclic graph, such that the child GO-term has more specific meanings than the parent GO-term. At present, GO is the only solution to describe the function of genes systematically and computationally and has become de facto standard for gene annotations [15, 16].

To measure tissue specificity of the functions by gene expression profiles, it is required to sum up the expression value of the genes along the hierarchical structure of gene ontology. Therefore, the expression values are required to be addable. The expression profiles also need to be measured cyclopedically. Among the gene expression-measuring methods, CAGE is the only one that satisfies these two conditions [17]. CAGE is a sequence based technology with which to collect 20bp tags from the 5'-end of transcripts in a sample exhaustively [9, 10, 17–19]. The collected tags are sequenced and mapped to the genome to identify which genes are expressed. As the number of the mapped tags represents the number of mRNA of the target gene in the sample, the gene expression data measured by CAGE is addable.

EST and SAGE also measure gene expression by sequencing transcripts. EST makes cDNA clones from transcripts and determines clone sequences from the 5'-end or 3'-end. The sequence lengths are several hundred bases. EST has contributed greatly to the detection of genes, gene annotations and establishment of the gene expression profiles. However, the sequences used as EST are too long to identify the expressed gene on a cost. Therefore, EST has been replaced with a more efficient method, namely SAGE. SAGE collects 10bp tags from the 3'-end of transcripts. As 10 to 20 tags are concatenated at a time to determine the sequence, SAGE is an efficient method to observe gene expression. However, SAGE requires the recognition sequence of NlaIII, namely CATG, near the 3'-end of the transcripts in order to collect the tags. It disturbs SAGE to observe the expression of genes exhaustively.

DNA microarrays and MPSS use competitive hybridization to observe gene expression. They anchor complementary strands of target genes on glass plates or solid beads, dye genes from the target condition and control condition with different colors, and then competitively hybridize the genes to the complementary strands. Gene expression is measured as the proportion of the brightness of the two colors. These methods measure the gene expression in the target condition based on the magnification factor over the control. Therefore, they cannot compare the expressions of two genes in the same condition. More directly saying, it is impossible to know which gene is more highly expressed. The gene expression data from these methods is not addable, and cannot be used to determine tissue-specific functions.

One of the simplest ways to measure the tissue-specificity of a function is the summation of the expression amount of the genes that play a role of the function. It can be reasonable, but includes an imperfection. The functions of house-keeping genes, which are expressed highly in all tissues and consequently they are inappropriate as tissue-specific, will be over estimated. To avoid the imperfection, we applied the information content that is an idea in the field of information theory [20] to our measurement of tissue-specificity.

Let E be an event that occurs according to the probability p, then information content of the event E is defined as $-\log(p)$. In this paper, the events correspond to the functions or GO-terms. The probability of a GO-term is defined as X/Y, where X is the summation of the CAGE tags of the genes that have the GO-term and Y is the summation of all the tags in the tissue. The information content of the GO-term is $-\log(X/Y)$. We defined the tissue-specificity of a GO-term as the difference in information content between the target tissue and whole tissues. We used 11,567,973 CAGE tags from 22 tissues of mouse offered by the FANTOM consortium [17] for the analysis of tissue-specific functions.

2. Results

Twenty two tissues of mouse CAGE data [17] were used to calculate tissue specific functions, and we selected liver, in which the largest number of tags are measured among 22 tissues, to show the results. GO-terms had three main categories: molecular function, biological process and cellular component. Due to space limitation, we chose molecular function and cellular component to show the result. As no usual method were utilize the countable expression data and the hierarchy of Gene Ontology, we show the Top 20 GO-terms of the most expressed genes for the comparison. The most expressed genes means that the numbers of mRNA in the

Table 1. Top20 TSFs of liver on Molecular Function

TSF	rank MEG	GO-term	EC number	#tag	#norm. tag
1	9139	biphenyl-2,3-diol 1,2-dioxygenase activity	1.13.11.39	3	0.9
2	4527	antifungal peptide activity		33	9.8
3	1672	endogenous peptide antigen binding		153	45.5
4	199	alcohol sulfotransferase activity	2.8.2.2	1214	361.3
5	120	tyrosine transaminase activity	2.6.1.5	1837	546.7
6	266	urocanate hydratase activity	4.2.1.49	941	280.0
7	140	betaine-homocysteine S-methyltransferase activity	2.1.1.5	2660	791.6
8	140	homocysteine S-methyltransferase activity	2.1.1.10	3030	901.7
9	103	urate oxidase activity	1.7.3.3	2219	660.3
9	-	oxidoreductase activity, acting on other nitrogenous compounds as donors		2219	660.3
9	-	oxidoreductase activity, acting on other nitrogenous compounds as donors, oxygen as acceptor		2219	660.3
12	14	4-hydroxyphenylpyruvate dioxygenase activity	1.13.11.27	7919	2356.6
12	14	quercetin 2,3-dioxygenase activity	1.13.11.24	7919	2356.6
14	369	plasmin activity	3.4.21.7	701	208.6
15	538	carbamoyl-phosphate synthase (ammonia) activity	6.3.4.16	513	152.7
15	538	carbamoyl-phosphate synthase (glutamine-hydrolyzing) activity	6.3.5.5	513	152.7
17	558	protein C (activated) activity	3.4.21.69	496	147.6
18	249	acyl-CoA ligase activity		992	295.2
18	249	butyrate-CoA ligase activity	6.2.1.2	992	295.2
20	275	triglyceride binding		907	269.9

cell are the largest among all the genes, which was a basic method that utilizes the countability to reveal the tissue specificity. The rankings of all the 22 tissues and the three GO-term categories are available from the certificated WEB page: *http://tsf.ics.es.osaka-u.ac.jp* with *(user,password) = (TSF,storia1441)*[a]. The supplementary tables are also placed the WEB page described above.

---

[a]Certification will be removed after the paper is accepted

Table 2.   Top20 MEGs of liver on Molecular Function

rank TSF	rank MEG	GO-term	EC number	#tag	#norm. tag
241	1	ferric iron binding		62181	18504.3
1136	2	chaperone activity		43354	12901.6
1198	2	GTPase activity	3.6.5.1-4	61246	18226.0
1342	2	structural molecule activity		93385	27790.2
1218	2	structural constituent of cytoskeleton		53137	15812.9
1089	2	GTP binding		88836	26436.5
1129	2	MHC class I protein binding		43004	12797.4
736	3	peroxidase activity	1.11.1.7	30531	9085.6
376	3	glutathione peroxidase activity	1.11.1.9	20721	6166.3
407	3	oxidoreductase activity		265408	78982.1
382	4	MHC class I receptor activity		24287	7227.5
170	5	monooxygenase activity		40237	11974.0
71	5	oxidoreductase activity, acting on paired donors, with incorporation or reduction of molecular oxygen, reduced flavin or flavoprotein as one donor, and incorporation of one atom of oxygen		26818	7980.7
649	6	transcription corepressor activity		16399	4880.1
1237	6	protein binding		291171	86648.8
454	7	carrier activity		115575	34393.7
505	7	lipid binding		49978	14872.8
166	8	electron carrier activity		21581	6422.2
162	8	oxidoreductase activity, acting on single donors with incorporation of molecular oxygen, incorporation of two atoms of oxygen		22954	6830.8
108	8	cysteine dioxygenase activity	1.13.11.20	10529	3133.3

2.1. *Molecular Functions*

Table 1 shows top 20 GO-terms of molecular function calculated by our method (TSF), and table 2 shows top 20 GO-terms of the most expressed genes (MEG). The last rank of TSF was 2173rd and the last rank of MEG was 10311st. The last rank of TSF was equal to the number of GO-terms that was appeared in liver, and the last rank of MEG meant the number of genes that expressed in liver. In the tables, the column named TSF and MEG represent the rank of GO-term that is shown in the column GO-term. When GO-terms are associated with Enzyme Commission number, we wrote them in the column EC number. The column named #tag shows the number of tags of the GO-term, which is calculated in **Step 2.** of our method. As the number of measured tags is different for each tissue, the column named #norm.tag shows the normalized number of tags by using a unit tpm (tags per million tags). There are "-" in the column of MEG rank. It represents that none of the genes with the GO-term expressed. The obvious differences between the two tables are #tag and the number of enzymes. The average #tag of TSF and MEG are 1874 and 69067, respectively. It represents that GO-terms in the top 20 TSF are not selected merely by the number of tags. The major part of the top 20 MEG rank was occupied by the common or abstract GO-terms, such as chaperone activity, GTPase activity, and structural molecule activity. They appeared in the MEG ranks of almost all tissues, and accordingly had lower TSF ranks. For example, ferric iron binding was ranked at 1st MEG, because *Trf*, whose gene description is transferin, was the most expressed gene in the liver. It also held 1st MEG rank at hippocampus, somatosensory cortex and visual cortex, and held 2nd to 10th MEG rank at 10 tissues. It was a typical function of the house keeping genes. There are 14 enzymes in the the TSF rank and four enzymes in the MEG rank. The frequent appearance of the enzymes in the top 20 TSF rank resulted from the intermediate metabolism that is one of the main roles of liver. Table 3 shows the number of enzymes that appeared in the Top20 TSF and MEG rank for each tissue. The average number of enzymes in the TSF rank was 5.3 and the average in the MEG rank was 2.9. In the table, liver and prostate grand had 14 enzymes in the TSF rank, which is the largest number among the 22 tissues, and followed by embryo and heart with 7 enzymes. To clarify the difference between liver and prostate grand, we categorized the enzymes according to the metabolic pathways in the KEGG (Kyoto Encyclopedia of Genes and Genomes) [21]. Nine of the 14 enzymes in the liver are responsible for metabolism, and of these, seven enzymes were involved in amino acid metabolism[b]. On the other hand, the prostate gland was rich in glycolipid metabolism. Five of the 14 enzymes were for glycolipid metabolism, two for carbon hydride metabolism, and two for amino acid metabolism. The result of our method well reflected the major role of liver, intermediate metabolism of amino acids.

2.2. *Cellular Component*

Table 4 shows top 20 TSF rank at liver on cellular component[c]. The last rank of TSF was 520th and the last rank of MEG was 9307th. In the table, three GO-terms appeared in both of top 20 TSF rank and top 20 MEG rank. They were 1) endocytotic vesicle (3rd TSF, 1st MEG), 2) peroxisome (6th TSF, 10th MEG) and 3) microsome (9th TSF, 5th MEG). The GO-terms in the both rank indicate that they were tissue specific and major transcripts in liver. In the followings, we validate

---

[b]The precise result is shown in Supplementary Table S4.
[c]the result of the top 20 MEG rank are available via our web site

Table 3. Number of Enzymes in Top20 TSF ranking and MEG ranking

tissue	num. enzymes		tissue	num. enzymes	
	TSF	MEG		TSF	MEG
adipose	6	3	lung	4	1
amnion	4	6	macrophage	2	5
brain	4	3	mammary gland	2	1
cerebellum	4	4	medulla oblongata	1	3
cerebral cortex	3	3	muscle	6	3
diencephalon	1	2	placenta	3	2
embryo	7	3	prostate gland	14	2
eye	3	2	somatosensory cortex	3	2
heart	7	4	striatal primordia	3	2
hippocampus	5	2	testis	6	3
liver	14	4	visual cortex	3	4

them. 1) Endocytotic vesicles are membrane-bound intracellular vesicles formed by invagination of the plasma membrane around an extracellular substance. As peroxisome and microsome are types of vesicles, we describe the role of peroxisome and microsome in the followings. 2) Peroxisome is a small, membrane-bound organelle that uses dioxygen to oxidize organic molecules, and contains enzymes that produce and others that degrade hydrogen peroxide. They fall under the term microbody, which holds the 6th TSF rank in Table 4, and were named after peroxidase-rich vesicles. Peroxidase held 3rd MEG rank in molecular function of liver (Table 2), indicating that peroxidase is highly produced in the liver. Table S6 shows the TSF ranks and number of normalized tags on peroxisome and peroxidase. From the table, peroxisome held a high TSF rank only in liver and also had the largest number of normalized tags in liver. The number of normalized tags in liver was three times greater than in heart, which had the second largest number of normalized tags. On the other hand, peroxidase in liver did not have particularly high TSF rank or number of normalized tags. Our result helped to find the fact that the presence of peroxisome is characteristic of liver, but peroxidase is not. 3) Microsomes are small vesicular particles containing high-density lipid. As mentioned above, the metabolism of lipid is a major role of the liver, and the mobilization and biosynthesis of triacylglycerol, a kind of lipid, were highly and locally occurred in liver. It indicates the GO-term microsome in liver were adequate to highly ranked.

The last two GO-terms we focused on were intercellular canaliculus (11th TSF) and connexon complex (18th TSF). These terms belongs to a GO-term named intercellular junction. Figure 1 shows a summary of intercellular junctions and related

Table 4. Top20 TSFs of liver on Cellular Component

rank TSF	MEG	Gene Ontology	#tag	#tag (tpm)
1	957	ornithine carbamoyltransferase complex	269	80.1
2	372	membrane attack complex	1295	385.4
3	1	endocytic vesicle	57969	17250.8
4	-	recombination nodule	4	1.2
4	7824	late recombination nodule	4	1.2
6	10	peroxisome	38807	11548.5
6	-	microbody	38807	11548.5
8	3110	glycine dehydrogenase complex (decarboxylating)	56	16.7
9	5	microsome	47059	14004.2
10	2219	vesicular fraction	47159	14033.9
11	669	intercellular canaliculus	1166	347.0
12	295	mitochondrial outer membrane translocase complex	797	237.2
13	8219	lateral element	3	0.9
14	922	citrate lyase complex	284	84.5
15	1926	glycine cleavage complex	174	51.8
16	1045	alpha-ketoglutarate dehydrogenase complex (sensu Eukaryota)	244	72.6
16	-	alpha-ketoglutarate dehydrogenase complex	244	72.6
18	362	connexon complex	1588	472.6
19	389	electron transfer flavoprotein complex (sensu Eukaryota)	1278	380.3
19	-	electron transfer flavoprotein complex	1278	380.3

GO-terms. The figure also shows the TSF rank of each GO-term in liver. The GO-term *Intercellular junction* has eight children terms, six of the eight terms had tags from liver, and two of the six and one descendant term held high TSF ranks. The three were *intercellular canaliculus* (11th TSF), *connexson complex* (19th TSF) and *gap junction* (27th TSF). On the other hand, the ranks of desmosome and tight junction were relatively low. A desmosome is a type of intercellular junction peculiar to epithelial cells, and a tight junction is found in epithelial cells and endothelial cells. The results match very well not only to the features of liver, but also to the feature of lung, heart, muscle and prostate gland.

Table 5 and Table S7 show the TSF ranks and number of normalized tags of the GO-terms in Figure 1 respectively. Lung had the highest TSF rank and largest

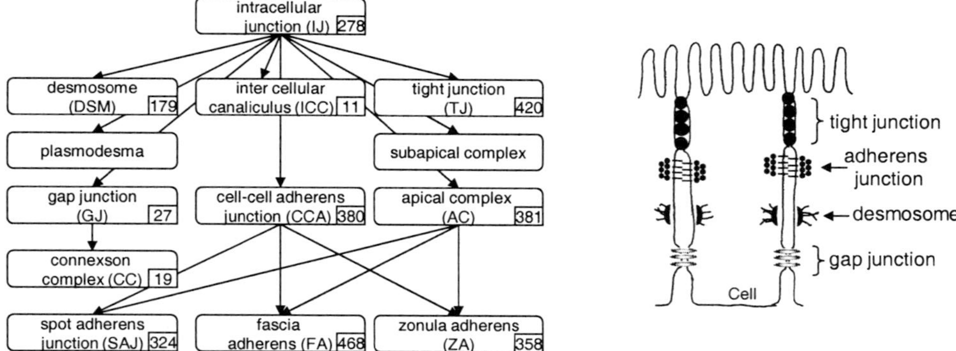

Fig. 1.   Intercellular Junction and related GO terms and thier TSF ranks in liver. The arrows represent the relation of parent and child between GO terms. The number at lower right corner of each GO term is the TSF ranks in liver. GO terms without the number has no tags. Abbreviation of the terms are used in table 4 and S10.

number of normalized tags for desmosome and tight junction. What interested us was the tradeoff relationships between liver and lung in TSF rank. If the TSF rank of liver was higher than 30th TSF, the TSF rank of lung became lower, and vice versa except for fascia adherens. For fascia adherens, both liver and lung had low TSF ranks, 468th and 403rd TSF rank respectively. On the other hand, muscle, heart and prostate gland had very high TSF ranks, 1st, 4th and 10th TSF rank respectively. A fascia adherens is a broad intercellular junction in the intercalated disk of cardiac muscle that anchors actin filaments. As the cardiac muscle is a component of heart, the high rank in heart seemed appropriate. The tags of fascia adherens in heart, muscle and prostate gland were derived from one gene, the nebulin-related anchoring protein. (MGI:1098765). It had six GO-terms, actin cytoskeleton organization and biogenesis, fascia adherens, myofibril, actin binding, metal ion binding and protein binding. Although prostate gland has no cardiac muscle, smooth muscle is developed at connective cells around the glandular cells.

3. Discussion

CAGE is one of the newest technologies to measure the expression of genes. The largest advantage of CAGE is that it enables us to compare the expression of the gene in question to that of other genes in the same tissue. To leverage the advantage, we put forward a scheme to sum up the gene expression value along the hierarchical structure of gene ontology, and proposed a method to reveal the tissue specific functions by calculating the differential information content. The results shown in the former section were well-adapted to the feature of liver, especially in the case of intercellular junctions. Generally speaking, many genes shares one GO-term, especially in the category of cellular component, in their annotations. This circumstance prohibits usual methods to exhibit tissue specific cellular component because they

Table 5. TSF ranks and normalized tags of intercellular junction and related GO terms.

tissue	IJ	CCA	GJ	TJ	DSM	AC	ICC	SAJ	ZA	FA	CC
					TSF rank						
adipose	443	434	439	413	465	420	472	-	394	128	412
brain	361	419	454	105	418	218	449	-	457	-	463
cerebellum	132	442	26	38	330	83	423	41	503	358	255
cerebral cortex	221	337	281	68	-	111	295	-	-	-	322
diencephalon	227	371	343	125	258	195	378	-	-	-	358
embryo	362	279	401	488	82	443	479	225	485	197	372
heart	204	167	201	291	35	158	237	-	287	4	182
hippocampus	143	200	-	38	-	90	-	-	-	-	-
liver	278	380	27	420	179	381	11	324	358	468	18
lung	23	21	426	13	29	14	227	25	8	403	378
macrophage	361	190	474	469	486	443	450	-	317	490	475
mammary gland	35	20	-	-	-	-	-	-	-	-	-
medulla oblongata	116	206	73	53	-	111	-	-	-	-	-
muscle	284	164	368	418	123	384	373	-	416	1	382
prostate gland	189	364	318	72	206	116	420	1	378	10	270
somatosensory cortex	396	460	162	225	465	365	410	-	496	358	169
striatal primordia	262	190	321	327	244	240	333	-	90	-	301
testis	210	151	101	409	342	357	447	-	195	-	68
visual cortex	390	464	237	244	442	355	442	-	484	-	242

Note: Abbreviations use in the table is same as figure. 1

do not take countability of expression in to account. There is a point to be improved
in our method. In the results, some GO-terms held same rank and they were often
parent-child in the GO hierarchy. They could be redundant and may disturb the
further analysis.

4. Conclusion

We presented a new method to describe tissue-specific function that leverages the
advantage of CAGE. The method used molecular function of Gene Ontology terms
to describe the tissue-specific functions and measure the tissue-specificities by the
Information Content of the terms. As a by-product of using Gene Ontology, it gave
us the information about tissue specificity on not only molecular functions, but
also biological process and cellular component. The method was applied to the

CAGE data on 22 tissues, and TSF ranks and also MEG ranks were calculated. The majority of the results agree well with the well-known tissue roles. The results of our method will provide researchers a clue to the further research of the tissue roles and the deeper functions of the tissue-specific genes.

5. Materials and Methods

5.1. *Materials*

Mouse CAGE tag data, which includes 11,567,973 tags, were used as the gene expression profiles. The tags were derived from 23 types of tissues. As the 23rd tissue is UNDEFINED TISSUE TYPE, we eliminated it from the analyses. Tags with the same sequence were clustered to a representative tag, and representative tags were mapped to mouse mm5 genes or, namely, FANTOM3's Transcriptional Unit (TU) [17]. The TUs were annotated with controlled nomenclature vocabulary transferred from the original literature and/or GO terms by the curators. We eliminated genes that have only one tag in total among all tissue. They are regarded as in the range of errors [22]. All data is available at the FANTOM3 Web page, and the details are described in the FANTOM3 papers [17].

5.2. *Methods*

The tissue specificities of GO-terms are measured using the differential information content (DIC). The GO-terms are ranked by DIC, and are called TSF rank. Let $Exp[tissue][gene]$ be the number of tags of *gene* at *tissue*, $GO[gene]$ be a set of GO-terms that *gene* has as its annotation. The TSF rank is calculated as follows.

Step 1. Calculate the expression amount of GO-terms *term* for each *tissue*:

$$GO\_Exp[term][tissue] = \sum_{term \in \text{Ancestors of } GO[gene]} Exp[tissue][gene].$$

Step 2. Sum up the expression amount of all tissues for each GO-term *term*:

$$GO\_Exp[term][all] = \sum_{\text{all tissues}}^{tissue} GO\_Exp[term][tissue].$$

Step 3. Calculate DIC of each *tissue* for each GO-term *term*:

$$DIC[term][tissue] = -\log \frac{GO\_EXP[term][tissue]}{GO\_EXP[root][tissue]} - \left(-\log \frac{GO\_EXP[term][all]}{GO\_EXP[root][all]} \right)$$

,where *root* is the root of GO, gene ontology.

Step 4. Rank GO-terms for each tissue according to descending order of DIC.

In the paper, we showd MEG ranks of GO-terms for each tissue. The MEG is ranked according to descending order of *MEG_SCORE*. It is calculated as follows:

$$MEG\_Score[term][tissue] = \sum_{term \in GO[gene]} Exp[tissue][gene].$$

References

[1] M. B. Eisen, P. T. Spellman, P. O. Brown and D. Botstein, *Proc. Natl. Acad. Sci. USA* **9**, 14863 (1998).

[2] t. Akutsu, S. Kuhara, O. Maruyama and S. Miyano, *Genome Informatics* **9**, 151 (1998).

[3] N. Friedman, M. Linial, I. Nachman and D. Pe'er, *Journal of Computational Biology* **7**, 601 (2000).

[4] P. T. Spellman, G. Sherlock, M. Q. Zhang, V. R. Iyer, K. Andres *et al.*, *Mol. Biol. Cell* **9**, 3273 (1998).

[5] M. D. Adams, J. M. Kelley, J. D. Gocayne, M. Dubnick, M. H. Polymeropoulos *et al.*, *Science* **252**, 1651 (1991).

[6] V. E. Velculescu, L. Zhang, B. Volgelsteine and K. W. Kinzler, *Science* **270**, 484 (1995).

[7] S. Saha, A. B. Sparks, C. Rago, V. Akmaev, C. J. Wang *et al.*, *Nat. Biotechnol.* **20**, 508 (2002).

[8] S. Brenner, M. Johnson, J. Bridgham, G. Golda, D. H. Lloyd *et al.*, *Nat. Biotechnol.* **18**, 630 (2000).

[9] T. Shiraki, S. Kondo, S. Katayama, K. Waki, T. Kasukawa *et al.*, *Proc. Natl. Acad. Sci. USA* **100**, 15776 (2003).

[10] R. Kodzius, Y. Matsumura, T. Kasukawa, K. Shimokawa, S. Fukuda *et al.*, *FEBS Lett.* **559**, 22 (2004).

[11] J. Schug, W. P. Schuller, C. Kappen, M. Salbaum, J. M. Bucan and C. J. Stoeckert Jr, *Genome Biology* **6:R33** (2005).

[12] K. Kadota, S. Nishimura, H. Bono, S. Nakamura, Y. Hayashizaki, Y. Okazaki and K. Takahashi, *Physiol Genomics* **12**, 251 (2003).

[13] K. Kadota, J. Ye, Y. Nakai, T. Terada and K. Shimizu, *BMC Bioinformatics* **7:294** (2006).

[14] Gene Ontology Consortium, *Genome Res.* **11**, 1425 (2001).

[15] Y. Okazaki, M. Furuno, T. Kasukawa, J. Adachi *et al.*, *Nature* **420**, 563 (2002).

[16] T. Imanishi, T. Itoh, Y. Suzuki *et al.*, *PLoS Biology* **2**, 856 (2003).

[17] P. Carninci, T. Kasukawa, S. Katayama *et al.*, *Science* **309**, 1559 (2005).

[18] P. Carninci, C. Kvam, A. Kitamura, T. Ohsumi, Y. Okazaki *et al.*, *Genomics* **37**, 327 (1996).

[19] P. Carninci and Y. Hayashizaki, *Methods Enzymol* **303**, 19 (1999).

[20] K. Ito (ed.), *Encyclopedic Dictionary of Mathematics* (The MIT Press, 1993), ch. 213. Information Theory, second edn.

[21] M. Kanehisa, *Science and Technology Japan* , 33 (1996).

[22] M. C. Firth, L. G. Wilming, A. Forrest, H. Kawaji *et al.*, *PLoS Genetics* **2** (2006).

FUNCTIONAL CENTRALITY: DETECTING LETHALITY OF PROTEINS IN PROTEIN INTERACTION NETWORKS

Kar Leong Tew[1]

kltew@i2r.a-star.edu.sg

Xiao-Li Li[1]

xlli@i2r.a-star.edu.sg

Soon-Heng Tan[1,2,3]

chris.tan@utoronto.ca

[1] *Knowledge Discovery Department, Institute for Infocomm Research, 21 Heng Mui Keng Terrace, Singapore 119613*

[2] *Samuel Lunenfeld Research Institute, Mount Sinai Hospital, Toronto, Ontario, Canada**

[3] *Department of Molecular and Medical Genetics, University of Toronto, Toronto, Ontario, Canada**

** Present Affiliation*

Abstract

Identifying lethal proteins is important for understanding the intricate mechanism governing life. Researchers have shown that the lethality of a protein can be computed based on its topological position in the protein-protein interaction (PPI) network. Performance of current approaches has been less than satisfactory as the lethality of a protein is a functional characteristic that cannot be determined solely by network topology. Furthermore, a significant number of lethal proteins have low connectivity in the interaction networks but are overlooked by most current methods.

Our work reveals that a protein's lethality correlates more strongly with its "functional centrality" than pure topological centrality. We define functional centrality as the topological centrality within a subnetwork of proteins with similar functions. Evaluation experiments on four *Saccharomyces cerevisiae* PPI datasets showed that NFC performed significantly better than all the other existing computational techniques. Our method was able to detect low connectivity lethal proteins that were previously undetected by conventional methods. The results and an online version of NFC is available at http://lethalproteins.i2r.a-star.edu.sg

Keywords: Lethal proteins; Functional centrality; Protein similarity; Protein-protein interaction

1. Introduction

A lethal (or essential) protein is one that renders the cell unviable on its removal. From a theoretical point of study, lethal proteins play an intricate role for cell survival and development. Studying of lethal proteins will open opportunities to understand other species and identification of potential drug targets [1]. While lethal proteins can be detected from gene knockout experiments, large-scale systematic detections can still be time-consuming and cost-prohibitive. As Jeong [2] noted, lethality profiles of a substantial number of genes are still unknown.

Alternative approaches to detect potential lethal proteins is thus required. One common hypothesis is that lethal proteins are strategically located within the protein-protein interaction (PPI) network such that their absence would create an adverse disruption to the topological stability of the network, thereby leading to biological lethality. Jeong [3] were one of the first to establish that there indeed exists a correlation between lethal proteins and their topological feature (connectivity) in the underlying PPI network. This led to a series of similar works unveiling new topological characteristics related to a protein's lethality (see Section 2).

However, the performance of many topological-based approaches had been less than satisfactory as the biological lethality of a protein is a functional characteristic that is unlikely to be adequately determined solely by network topology. Moreover, many current approaches were based on the assumption that a protein's lethality is correlated with high connectivity in the PPI network. This may not be always true as it is possible for a low-connectivity protein to be lethal. We found a substantial number of known lethal proteins with low connectivity (number of interaction partners ≤ 5) in the yeast PPI network (see Table 1).

In this work, we combine the topological-based concept for protein lethality with the notion of functional modules [4, 5], which are groups of interconnected proteins performing discrete functions in the PPI network. Multimeric protein complexes (such as the ribosome that synthesize polypeptides from amino acids) and biological pathways are instances of functional module. We reasoned that lethal proteins are the key players or coordinators within functional modules and their removal will maximally disrupt the operations of the modules which impact cell fitness. We hypothesized that these key proteins should also be centrally positioned within functional modules to carry out their roles effectively and their removal will cripple the modules more easily than the removal of proteins lying at the peripheral.

Thus in this paper, we introduce a novel *neighborhood functional centrality* (NFC) measure to quantify the extent in which a protein is surrounded by functionally consistent neighboring proteins in the PPI network. We also devised a Neighborhood Functional Centrality (NFC) algorithm to mine lethal proteins in PPI networks based on this concept. Evaluation on four *Saccharomyces cerevisiae* PPI datasets showed that NFC performed significantly better than all other existing computational techniques. Given that many lethal proteins can be of low connectivity, we also verified that our NFC method can detect low-connectivity lethal proteins undetected by conventional methods.

2. Related Works

Jeong [3] first reported that the lethality of a protein is positively correlated to its connectivity (or degree) in the protein interaction network—the number of interacting partners a protein has. This has led to numerous subsequent works that attempted to infer a protein's lethality in baker's yeast using various other network topological characteristics such as clustering coefficient [6], betweenness [7], damage [8], and subgraph centrality [9]. The clustering coefficient quantifies the probability of two interacting proteins are also interacting with a similar third protein. The betweenness score quantifies a protein's topological centrality based on number of shortest paths that pass through it in the underlying PPI network. The damage score measures the disintegration of the underlying PPI network resulting from the removal of a protein. The subgraph centrality score quantifies the number of subgraphs a node participate in with emphasis on shorter closed paths.

All the above methods used solely topological measures that are directly or indirectly dependent on the high connectivity of proteins within the PPI network. As such, they will not work very well when the underlying PPI network is a sparse network. In the four PPI datasets that we have used for our evaluation experiments, we found an average of 67.0% of lethal proteins exhibited low connectivity (number of interaction partners ≤ 5) in the underlying PPI networks (see the bracketed figures in Table 1). Furthermore, a substantial amount (54.7%) of the high-connectivity proteins (number of interaction partners ≥ 6) were not known to be lethal, suggesting that the biological lethality of a protein cannot be adequately determined solely by network topology. In this paper, we propose a new method that incorporates functional information with topological information to better detect lethal proteins, including those with low connectivity in PPI networks.

3. Method

We model the protein interaction data as a large undirected graph $G_{PPI} = (V_{PPI}, E_{PPI})$, where V_{PPI} represents the set of interacting proteins and E_{PPI} denotes all detected pairwise interactions between two proteins from V_{PPI}. Our NFC algorithm consists of two steps. First, for each protein in the interaction graph, we construct a local neighborhood graph to compute a *nfc* score (Section 3.1). Then, we assess the significance of $nfc(u)$ by computing its corresponding Z_{nfc} (Section 3.2).

3.1. *Computing the Neighborhood Functional Centrality*

To compute the neighborhood functional centrality score *nfc* for each protein in the interactome, we define the neighborhood graph for each vertex u in G_{PPI} as follows:

Definition 1.   For each vertex $u \in V_{PPI}$, its neighborhood graph is defined as $G_u = (V_u, E_u)$, where:

$$V_u = \{v \mid v \in V_{PPI} \wedge dist(u, v) \leq \theta\},$$

$$E_u = \{(v_j, v_k) \mid (v_j, v_k) \in E_{PPI} \wedge v_j, v_k \in V_u\}, \text{ and}$$

$dist(u, v)$ is a function that returns the shortest distance between u and v.

The neighborhood graph G_u of a vertex u is the subgraph in G_{PPI} induced by the vertices that are within a radius of θ from u. θ is a user-defined variable to control the radius (or size) of the neighborhood graph of vertex u and we will investigate its effect on the prediction results later (Section 5.4).

Next, we evaluate whether a protein is functionally central in its neighborhood graph. This involves measuring the functional similarities among the proteins in the neighborhood graph. This is achieved by incorporating functional information associated with each protein into our analysis.

Biological functions are typically organized in a hierarchical structure—generic biological functions (such as *transcription*) can be progressively broken down into more specific functions (such as *transcription termination*, and *transcription from RNA polymerase II promoter*). Each protein in an interactome is annotated (if at all[a]) with functions at various levels of specificity depending on the state of functional knowledge on the individual proteins. Currently, the most commonly used structure for functional annotation is the Gene Ontology—GO [10].

To compute the functional centrality of the proteins, we take into consideration that the proteins' functional annotations are in ancestor/descendent relationships. As such, we adopted the Relative Specificity Similarity (RSS) method that Wu [11] have developed which is a quantitative measure of the similarity between two GO functions (Definition 2) taking into account the hierarchical structure of GO:

Definition 2. Relative Specificity Similarity (RSS)

$$RSS(term_i, term_j) = \frac{maxDepth^{GO}}{maxDepth^{GO} + \gamma} \cdot \frac{\alpha}{\alpha + \beta}$$

where $maxDepth^{GO}$ is the maximum depth of the GO, α measures the maximum number of common ancestor terms shared between $term_i$ and $term_j$ in a single path, β is the value of the longer distance between $term_i$ and $term_j$ to their closest leaf nodes, and γ measures the shortest distance between $term_i$ and $term_j$. Refer to Wu [11] for details.

Definition 2 defines the functional similarity between two individual functions. However, a protein could be involved in different biological processes and associated with multiple GO annotations. Suppose F_u and F_v are the function annotations of proteins u and v respectively, we define the functional similarity between the proteins u and v as follows:

Definition 3. The protein functional similarity between two proteins u and v is defined as

$$protein\_funsim(u, v) = \frac{\sum_{i=1}^{|F_u|} \sum_{j=1}^{|F_v|} RSS(F_{(u,i)}, F_{(v,j)})}{(|F_u| * |F_v|) * dist(u, v)}$$

where $F_{(u,i)}$ and $F_{(v,j)}$ denote protein u's i-th and protein v's j-th's functions respectively, and $|F_u|$ denotes the number of functions protein u is annotated with.

---

[a]We will discuss strategies to handle proteins without functional annotations and their prediction in Section 5.3.

Definition 3 quantifies the extend of functional similarity between two proteins which may have multiple functions. The denominator $dist(u, v)$ is included here to give higher weightage for protein pairs that are closer together in the underlying interaction graph—this takes into account the implicit functional similarity between the two proteins based on their distance in the interactome.

We are now ready to define the *neighborhood functional centrality* score for each protein based on its functional similarity with proteins in its neighborhood graph:

Definition 4. The neighborhood functional centrality $nfc(u)$ of a protein u is defined as

$$nfc(u) = \sum_{v \in V_u, v \neq u} protein\_funsim(u, v)$$

Definition 4 quantitates the degree of functional consistency between protein u and all the other proteins in its neighborhood graph $G_u = (V_u, E_u)$. The value $nfc(u)$ indicates the functional centrality of protein u in G_u.

3.2. *Computing the Corresponding Z-scores*

Depending on the underlying functional distribution of the proteins in the interactome, it is possible that protein u is more likely to be assigned a higher $nfc(u)$ when located in a larger neighborhood graph G_u, or vice versa due to the summation used in Definition 4. In other words, given an interactome, the statistical distributions of $nfc(u)$ in smaller neighborhoods may be different from those in bigger neighborhoods (i.e. different means and/or different standard deviations). The significance of a particular $nfc(u)$ value is therefore dependent on the underlying distribution with respect to the size of the local neighborhood chosen for u.

In this work, we assess the significance of each protein's $nfc(u)$ value by computing its Z-Score (or "standard scores") Z_{nfc} as follows:

Definition 5. $Z_{nfc}(u, s)$ is defined as

$$Z_{nfc}(u, s) = \frac{nfc(u) - \mu_s}{\sigma_s}$$

where μ_s and σ_s are the mean and standard deviation of the distribution of $nfc(u)$ values computed from neighborhood graphs of size s.

Definition 5 requires computation of the distributions of neighborhood functional centrality values for differently-sized neighborhood regions. In fact, we only need to compute distributions for neighborhood sizes actually used in our $nfc(u)$ computation which we stored in set US. We estimate the distributions of neighborhood functional centrality values for each neighborhood size in US by randomly fetching same-sized neighborhood graphs for each vertex (if possible[b]) to determine

---

[b]It is possible that we are unable to fetch from u, a neighborhood of an intended size if the protein is in a small isolated partition. However, θ can be set small enough such that it is possible to find some neighborhoods of the intended size with some other vertices in the PPI network.

the corresponding neighborhood functional centrality values.

4. Experimental Data

For evaluation, we performed comparative experiments to show that our neighborhood functional centrality (NFC) approach performs better than other existing computational techniques. We used PPI datasets for *Saccharomyces cerevisiae* as it is currently the only organism with fairly complete knockout analysis (which forms our core lethal protein list).

4.1. *PPI Datasets*

We used four publicly available *Saccharomyces cerevisiae* protein interaction datasets for our evaluation experiments: *FYI* [12], *Nature* [2], *Bu* [13] and *DIPS* [14]. Each dataset was named after the source from which we have acquired them—details about each dataset are shown in Table 1. We have elected to use four different datasets so as (a) to facilitate direct comparisons with previous work and (b) to verify the performance against datasets of varying quality. The first dataset *FYI* is a high-quality (reliable) but sparse yeast interaction dataset with minimal false positives [12]. Another sparse network is *Nature*—included as it was employed by Jeong [3] whom first used the connectivity measure (which we will be comparing against) to detect the lethal proteins. The third dataset *Bu* is a relatively dense network with 3 times as many interactions as the previous two datasets. It was compiled by Bu [13] for function prediction, and subsequently used by Estrada [9] whom introduced the Subgraph Centrality (SC) measure which we will also be comparing against. The fourth dataset *DIPS* was obtained from the Database of Interaction Proteins (Nov 2005), giving rise to another dense network with interactions derived from various biological experiments. We pre-processed all four datasets by removing self-interacting interactions and isolated protein pairs from the networks.

Table 1.   Details of the four *Saccharomyces cerevisiae* protein interaction datasets used in our evaluation experiments.

	FYI	**Nature**	**Bu**	**DIPS**
# Proteins	1210 (*958*)	1638 (*1490*)	2224 (*1531*)	2406 (*1773*)
# Lethal	464 (*333*)	369 (*312*)	670 (*349*)	695 (*414*)
# Unknown (No Function)	12 (*10*)	94 (*84*)	18 (*17*)	23 (*23*)
# Interactions	2400	2201	6609	5665

Note: Italicized numbers in brackets represents proteins with connectivity ≤ 5.

Since our NFC method incorporates the functional information of the proteins for evaluation, we used function annotations classified as biological process by GO [10] (27-Oct-2006). Functional annotation has not yet reach the stage where we can expect all the proteins to be annotated (see "# Unknown" in Table 1) and we address this and the function prediction mechanisms in Section 5.3.

4.2. *Reference List and Evaluation Metric*

For evaluation, we used a benchmark lethal protein list (the Core list) consisting of 1106 known lethal proteins for *Saccharomyces cerevisiae* determined by PCR-based gene deletion strategy [15]. This set of lethal proteins was derived experimentally using PCR-based gene deletion strategy [16, 17]. We plot the corresponding ROC (Receiver Operating Characteristic) curves to compare the performance of the various prediction methods. Quantification of the significance of each prediction technique's ROC curve is done using the AUC (Area Under the Curve) values.

5. Experimental Results

In this section, we first compare our NFC method against other existing methods for predicting lethal proteins from PPI datasets to see whether NFC can perform better than the current methods (Section 5.1). We also check on the performance of our NFC to see if it can better detect low connectivity lethal proteins (Section 5.2).

We next investigate the performance of NFC in the absence of functional information and how function prediction mechanism can help in addressing this issue (Section 5.3). Finally, we investigate how the performance of NFC may be affected by different values of θ which controls the neighborhood radius (Section 5.4).

5.1. *Performance Comparisons*

We compare the performance of NFC against three other existing methods, namely, connectivity [3], subgraph centrality (SC) [9] and cluster coefficient (CC) [6]. For a fair evaluation, we use the same four protein interaction datasets, core lethal protein list, and function annotation for all the methods. We have omitted here the *damage score* method proposed by Schmith [8] and the *betweenness score* method by Joy [7]. This is because the damage score was already known to have a lower correlation to lethality as opposed to connectivity in PPI datasets [8], while betweenness have been outperformed by SC [9].

Table 2.   AUC comparisons of NFC, Connectivity, SC, and CC.

	FYI	**Nature**	**Bu**	**DIPS**
NFC	67.8 (*67.7*)	71.2 (*73.2*)	74.9 (*72.0*)	75.3 (*74.3*)
Connectivity	60.8 (*58.1*)	61.0 (*58.6*)	66.0 (*58.3*)	65.8 (*60.9*)
SC	57.1 (*54.2*)	56.8 (*53.4*)	65.4 (*56.7*)	63.9 (*58.7*)
CC	55.2 (*56.0*)	59.0 (*59.1*)	58.8 (*59.1*)	56.9 (*58.9*)

Note: Italicized numbers in brackets represents AUC values for detecting proteins with connectivity ≤ 5.

In Figure 1, we show the ROC curves of the four prediction methods on our experimental datasets. The AUC values for all four datasets are shown in Table 2 which clearly depicts the generality of NFC when used in datasets of varying size and quality. The results also shows that NFC can better detect lethal proteins from PPI datasets than other existing techniques due to its larger AUC values.

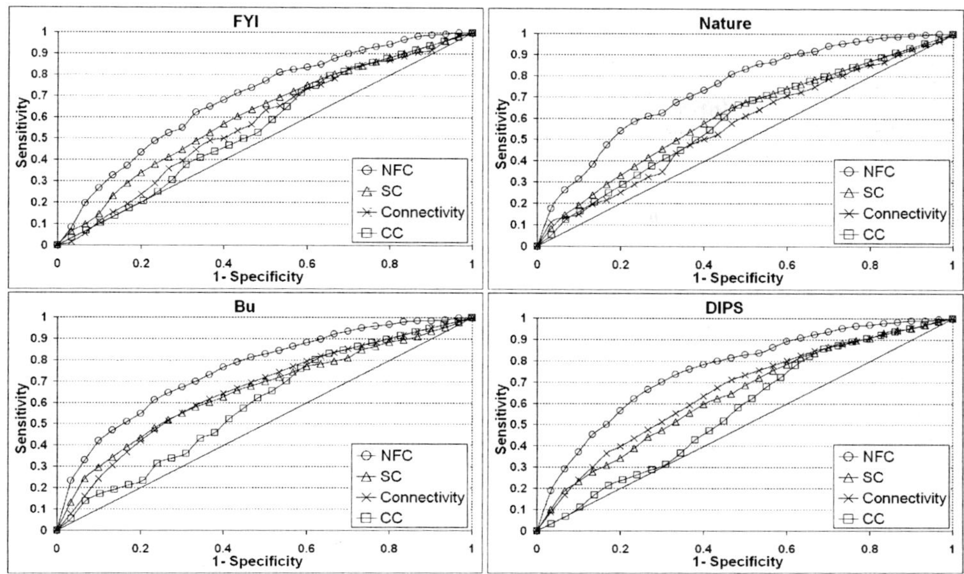

Fig. 1.   ROC charts of NFC, Connectivity, SC, and CC for each evaluation dataset.

5.2. *Low Connectivity Proteins*

As from Table 1, a significantly large proportion (78.2% on average) of proteins in the datasets are of low-connectivity (i.e. number of interaction partners ≤ 5), even in dense PPI networks such as Bu and DIPS. Furthermore, a high average of 67.0% of the lethal proteins in our core reference list also has low connectivity in the underlying PPI networks. This means that the connectivity-based detection methods would have missed out a significant number of lethal proteins if we rely solely on detecting high-connectivity proteins. The bracketed numbers in Table 2 confirms that NFC can detect low connectivity lethal proteins much better than the other existing methods in all four datasets.

5.3. *Protein Function Annotation: Absence and Prediction*

The incorporation of biological knowledge in addition to topological information have vastly improved lethal proteins detection. However, this implies that our NFC method is dependent on the amount of biological knowledge available, and its performance is expected to decrease with a higher number of unknown proteins (i.e. proteins without known functions).

We tested this on the datasets by generating the situation where 50% of the proteins have an unknown function through a random selection process of marking a protein as having unknown function[c]. As expected, a decline was observed in the

---

[c] Here, we set the upper limit at 50% as a statistical study shows the largest percentage of unknown proteins on other species was 46.0% (*Caenorhabditis elegans*).

AUC values. We then follow this up with the utilization of function prediction mechanism. For simplicity, we chose the *Majority* measure proposed by Schwikowski [18]. By using the *Majority* method in the same situation (50% unknown), the improvements in AUC values for each dataset are from 54.8% to 62.5% for FYI, 57.4% to 62.2% for Nature, 60.0% to 69.1% for Bu, and 58.8% to 69.7% for DIPS (Table 3). Even with a basic method, NFC is still able to obtain AUC values better than existing methods. By coupling with more sophisticated protein function prediction methods [19–21], we certainly expect NFC performance to be more robust than illustrated.

Table 3.   AUC values with different percentages of unknown proteins.

	Normal	10%	20%	30%	40%	50%
FYI	67.8	64.5 (*67.8*)	60.7 (*66.8*)	58.7 (*65.4*)	56.6 (*64.2*)	54.8 (*62.5*)
Nature	71.2	68.4 (*70.6*)	64.6 (*68.8*)	61.7 (*66.9*)	59.9 (*64.8*)	57.4 (*62.2*)
Bu	74.9	70.9 (*73.8*)	67.1 (*72.4*)	63.4 (*71.5*)	61.0 (*70.4*)	60.0 (*69.1*)
DIPS	75.3	70.9 (*74.6*)	66.9 (*73.4*)	64.3 (*72.2*)	60.9 (*70.9*)	58.8 (*69.7*)

Note: Italicized numbers in brackets represents AUC values with the *Majority* measure used.

5.4. *Varying the Neighborhood Radius Threshold θ*

Recall that our NFC method employed a user-defined threshold θ (Section 3.1) that controls the radius of the neighborhood graphs to compute the functional centrality values[d]. It is therefore possible that the performance of NFC may be affected by choice of θ used. Further evaluation experiments where we computed the various AUC values for each dataset when $\theta = 1$ to $\theta = 5$ has the mean deviation of 0.3% (FYI), 1.2% (Nature), 1.1% (Bu), and 1.3% (DIPS) (Table 4). These values are a clear indication that NFC's performance is not affected by θ. During our investigation, we also found that by using the Z-score instead of the raw $nfc(u)$ values, our NFC method has effectively adjusted for the effects of different neighborhood sizes and improved the accuracy of its predicted lethal proteins. Compared to using only the raw $nfc(u)$ values in the computation, Z_{nfc} improved the AUC values by 6.2%, 7.1%, 5.8%, and 7.3% for FYI, Nature, Bu, and DIPS respectively.

Table 4.   AUC values for different θ.

	$\theta = 1$	$\theta = 2$	$\theta = 3$	$\theta = 4$	$\theta = 5$	Mean Deviation
FYI	67.0%	67.8%	67.1%	66.6%	66.6%	0.3%
Nature	67.4%	71.2%	70.8%	69.4%	68.5%	1.2%
Bu	72.2%	74.9%	73.5%	72.0%	71.7%	1.1%
DIPS	73.2%	75.3%	73.6%	71.9%	70.9%	1.3%

---

[d]Results presented above are obtained with $\theta = 2$.

6. Discussions and Conclusions

The detection of lethal proteins is useful for various aspect of biological study. To complement the costly experimental approaches such as PCR-based gene deletion strategy [15], and to exploit the large datasets of protein-protein interactions that have become available, researchers have proposed numerous computational methods using topological properties associated with high connectivity to infer protein lethality. However, we have shown in this paper that the lethality of a protein is a functional characteristic that cannot be determined solely by network topology. Furthermore, a significant number of lethal proteins have been found to have low connectivity (less than 5 interaction partners) in the interaction networks.

A protein's lethality should also be determined using additional non-topological information such as its functional grouping within the cell. We reasoned proteins that are the key players or coordinators within functional modules are likely to be lethal as their removal will drastically disrupt the effective operations of the modules. In this paper, we proposed a novel *neighborhood functional centrality* (NFC) approach that incorporated the conventional topological concept for protein lethality with the notion of functional modules [4, 5] to better detect protein lethality.

NFC was shown to discover both lethal proteins with high connectivity as well as those with low connectivity. In the top 100 lethal proteins detected by NFC from the FYI dataset, 27 bind to 5 or less proteins (low connectivity), 40 bind between 6 to 9 proteins, and 33 bind to 10 or more proteins. On average, NFC was able to detect three times more low-connectivity lethal proteins within the top 100 positions as compared to the connectivity method [3].

A functional distribution analysis of the top ranking lethal proteins reveal that NFC favors lethal proteins involved in basal cell activities. For example, the top 100 lethal proteins detected by NFC in each of the four datasets have GO functions that can be broadly grouped under "translation", "replication", and "transcription" categories. In contrast, the connectivity method by Jeong [3] favored the discovery of lethal proteins with "mitotic cell cycle and cell cycle control" and "fungal and other eukaryotic cell type differentiation" functions. Our preliminary take on the differences is that NFC's functional centrality assumption led to the tendency to find the cores of protein complexes common in some biological pathways, whereas the connectivity method favored the discovery of lethal proteins associated with different functions because such lethal proteins would need to interact with multiple proteins in order to coordinate the global cellular activities needed for cell growth and differentiation.

When a protein has functional annotations, an intelligent guess may be made with regards to its lethality based on the biological understanding of its annotated functions. For example, we would expect many proteins involved in translation to be lethal as the process is a basal cellular activity. However, only 12.7% of all the proteins with translation function are actually lethal. This could stem from our current incomplete understanding of the exact roles played by each protein in translation. On average, GO terms identified in our top 100 NFC proteins are found to associate with lethal proteins 27.3% of the time. On the other hand, 70.0% of the top 100 NFC are lethal where we made use of functional consistency between proteins rather than functional understanding. Thus, integrating PPI network with functional grouping of proteins enable us to better detect lethal proteins than just using functional information alone.

Interestingly, we also found 12 (FYI), 13 (Nature), 18 (Bu), and 12 (DIPS) instances of high-confidence (top 100) predicted lethal proteins that are not in the current reference lethal protein lists, but each has at least one homologous sequence (BLAST's *e-value* $\leq 1e^{-99}$). The presence of homologous copies of a protein within the same genome could potentially buffer the protein deletion which would otherwise lead to lethality. It is conceivable that these predicted proteins require the removal of its associated homologous for lethality to take effect.

Given that the core lethal protein set we used is an incomplete reference list, those highly ranked non-lethal proteins could be novel lethal proteins. We found numerous high-ranked proteins by NFC that turned out to be true lethal proteins listed in other lethal protein reference sets. For example, the proteins YLR268W and YFL017W-A, respectively ranked at the top 16th and 38th positions by NFC in the DIPS and Bu datasets, were absent from our core lethal protein list but found in another lethal protein list used by Jeong [3]. Further comparison of NFC's predictions with two other reference sets used by Jeong [3], and list compiled by *MIPS* [22], found that out of the top 500 ranked proteins, an additional 15 (FYI), 11(Nature), 18 (Bu) and 16 (DIPS) were recorded in these alternative lethal sets.

Regardless of the improved accuracy of our predictive models over time, biological validation of predictions is always necessary. Our hope is that the predictions from this and the future works on computational lethal protein detection can become a useful tool for focusing further experiments that can lead to a shorter time frame required for lethal protein discovery and understanding.

Acknowledgement

We would like to thank our colleagues See-Kiong Ng, Zeyar Aung, and Suryani Lukman for their invaluable assistance rendered during this project.

References

[1] Rosamond, J., and Allsop, A., Harnessing the power of the genome in the search for new antibiotics, *Science*, 287:1973–1976, 2000.

[2] Jeong, H., and Oltvai, Z.N., and Barabsi, A.L., Prediction of Protein Essentiality Based on Genomic Data, *Complexus*, 1(12):19–28, 2003.

[3] Jeong, H., Mason, S.P., Barabsi, A.L., and Oltvai, Z.N., Lethality and centrality in protein networks, *Nature*, 411(6833):41–42, May 2001.

[4] Hartwell, L.H., Hopfield, J.J., Leibler, S., and Murray., A.W., From molecular to modular cell biology, *Nature*, 402(6761):47–52, 1999.

[5] Spirin, V., and Mirny, L. A., Protein complexes and functional modules in molecular networks, *Proc Natl Acad Sci USA*, 100(21):12123–12128, Oct 2003.

[6] Yu, H., Greenbaum, D., Lu, H. X., Zhu, X., and Gerstein, M., Genomic analysis of essentiality within protein networks, *Trends Genet*, 20(6):227–231, 2004.

[7] Joy, M. P., Brock, A., Ingber, D. E., and Huang, S., High-betweenness proteins in the yeast protein interaction network, *J Biomed Biotechnol*, 2005(2):96–103, 2005.

[8] Schmith, J., Lemke, N., Mombach, J. C. M., Benelli, P., Barcellos, C. K., and Bedin, G. B., Damage, connectivity and essentiality in protein-protein interaction networks, *Physica A Statistical Mechanics and its Applications*, 349(3-4):675–684, Apr 2005.

[9] Estrada, E., Virtual identification of essential proteins within the protein interaction network of yeast, *Proteomics*, 6(1):35–40, Jan 2006.

[10] Ashburner, M., Ball, C. A., Blake, J. A., Botstein, D., Butler, H., *et al.*, Gene Ontology: tool for the unification of biology, *Nature Genet*, 25:25–29, May 2000.

[11] Wu, X., Zhu, L., Guo, J., Zhang, D. Y., and Lin, K., Prediction of yeast protein-protein interaction network: insights from the Gene Ontology and annotations, *Nucl. Acids Res.*, 34:2137–2150, 2006.

[12] Han, J. D. J., Bertin, N., Hao, T., Goldberg, D. S., Berriz, G. F., Zhang, L. V., *et al.*, Evidence for dynamically organized modularity in the yeast protein-protein interaction network, *Nature*, 430(6995):88–93, Jul 2004.

[13] Bu, D., Zhao, Yi., Cai, L., Xue, Hong., Zhu, X., Lu, H., *et al.*, Topological structure analysis of the protein-protein interaction network in budding yeast, *Nucl. Acids Res.*, 31(9):2443–2450, May 2003.

[14] Xenarios, I., Rice, D. W., Salwinski, L., Baron, M. K., Marcotte, E. M., and Eisenberg, D., Dip: the database of interacting proteins, *Nucl. Acids Res.*, 28(1):289–291, Jan 2000.

[15] Giaever, G., Chu, A. M., Ni, L., Connelly, C., *et al.*, Functional profiling of the *Saccharomyces cerevisiae* genome, *Nature*, 418(6896):387–391, Jul 2002.

[16] Baudin, A., Ozier-Kalogeropoulos, O., Denouel, A., Lacroute, F., and Cullin, C., A simple and efficient method for direct gene deletion in *Saccharomyces cerevisiae*, *Nucl. Acids Res.*, 21(14):3329–3330, July 1993.

[17] Wach, A., Brachat, A., Pohlmann, R., and Philippsen, P., New heterologous modules for classical or PCR-based gene disruptions in *Saccharomyces cerevisiae*, *Yeast*, 10:1793–1808, 1994.

[18] Schwikowski, B., Uetz, P., and Fields, S., A network of protein-protein interactions in yeast. *Nat Biotechnol*, 18(12):1257–1261, Dec 2000.

[19] Nabieva, E., Jim, K., Agarwal, A., Chazelle, B., and Singh, M., Whole-proteome prediction of protein function via graph-theoretic analysis of interaction maps, *Bioinformatics*, 21:302–310, 2005.

[20] Jiang, T., and Keating., A. E., AVID: An integrative framework for discovering functional relationships among proteins, *BMC Bioinformatics*, 6:136, 2005.

[21] Chua, H. N., Sung, W. K., and Wong, L., Exploiting indirect neighbors and topological weight to predict protein function from protein-protein interactions, *Bioinformatics*, 22:1623–1630, 2006.

[22] Guldener, U., Mnsterktter, M., Kastenmller, G., Strack, N., Helden, J. V., *et al.*, CYGD: the Comprehensive Yeast Genome Database, *Nucl. Acids Res.*, 33:D364–D368, 2005.

THE IN SILICO PREDICTION OF PROMOTERS
IN BACTERIAL GENOMES

MICHAEL TOWSEY[1]   JAMES M. HOGAN[2]
m.towsey@qut.edu.au   j.hogan@qut.edu.au

SARAH MATHEWS[1]   PETER TIMMS[1]
s.mathews@qut.edu.au   p.timms@qut.edu.au

[1] *Institute for Health and Biomedical Innovation, Queensland University of Technology, Queensland, Australia.*
[2] *Faculty of Information Technology, Queensland University of Technology, Queensland, Australia.*

In silico approaches to the identification of bacterial promoters are hampered by poor conservation of their characteristic binding sites. This suggests that the usual position weight matrix models of bacterial promoters are incomplete. A number of methods have been used to overcome this inadequacy, one of which is to incorporate structural properties of DNA. In this paper we describe an extension of the promoter description to include SIDD (stress induced duplex destabilization), DNA curvature and stacking energy. Although we report the best result to date for a realistic promoter prediction task, surprisingly, DNA structural properties did not contribute significantly to this result. We also demonstrate for the first time, that sigma-54 promoters have a stronger association with SIDD than do other promoter types.

Keywords: bacterial promoters, promoter prediction, SIDD, duplex destabilization, DNA curvature

1. Introduction

The identification of promoters is essential for an understanding of gene regulation. However wet-lab techniques to identify bacterial promoters are costly and time consuming and thus *in silico* methods have strong appeal. Unfortunately, computational approaches to promoter identification are confounded by the poor conservation of their important functional sites. Transcription in bacteria is initiated by a protein complex known as RNA polymerase (RNAP), consisting of five subunits (collectively known as the *core enzyme*) and an additional σ factor. The σ factor is responsible for locating promoters by recognizing two binding sites, typically located at the -10 and -35 positions with respect to the transcription start site (TSS). Once transcription has begun, the σ factor dissociates and transcription continues with core enzyme alone.

In silico identification of promoters has tended to focus on detecting the -10 and -35 binding site motifs which are typically (in the case of the most common housekeeping σ^{70}) separated by a spacer of 14 to 20 base pairs (bp). However, it was established early using information theoretic reasoning, that the known -35 and -10 binding sites are insufficiently conserved to account for all the expected promoters in the background genome [15]. Furthermore, when potential binding sites are scored using position weight matrices (PWMs), it is found that about 50% of the known TSSs are not located at the highest scoring position upstream of a gene start site [7]. Clearly there are other factors involved in the positioning of promoters that are not captured in a simple PWM description. In this regard, it is extremely interesting that experiments have demonstrated that many look-alike promoter sites initiate transcription *in vitro* even though they fail to do so *in vivo* [8].

Several attempts have been made to use more sophisticated machine learning methods to identify promoters, for example neural networks [3] and support vector machines (SVM) [4]. While these methods offer somewhat increased accuracy depending on how the task is constructed, the improvements may not justify the heavy computation required for training the classifiers. Maetschke *et al.* [11] revisited the PWM approach, but this time utilizing information that has recently come to light about the mode of action of RNA polymerase. Incorporating extended -10 motifs [12] and UP elements (AT rich regions upstream of the promoter) [18] into their promoter description slightly improved predictive accuracy to around 50%, but clearly the most important predictive improvement for *E. coli* promoters was obtained by including information about the distance of the putative TSS from the gene start site (hereafter referred to as the TSS-GSS distance). This observation has also been made in [3].

There are at least three explanations advanced to explain the poor predictive performance of existing bacterial promoter models. First, it is possible that potentially strong promoter sites are masked by some mechanism that makes them inaccessible to RNA polymerase. In chlamydial species, for example, DNA is condensed by the binding of histone-like proteins during late development, which plays a role in down-regulating gene expression by removing the accessibility of promoters [6]. Secondly it is well known that some weak promoters can only function in conjunction with activators. Unfortunately, while many transcription factors have been identified, most of their binding sites have not, and it is not clear how transcription factor binding sites can be included in promoter models, except in the case of some well characterized global regulators [16]. Thirdly, it is becoming increasingly clear that structural features of DNA have an important regulatory role in gene expression, for example stacking energy [1], DNA curvature [9, 13] and Stress Induced Duplex Destabilization (SIDD) [19]. Our paper investigates the use of these DNA structural properties to help identify promoters.

Stacking energy refers to the interactions between consecutive base pairs of a 'stacked' DNA sequence. It is assumed to be a purely local phenomenon depending only on nearest neighbour interactions and contributes to local duplex stability or meltability. Units are kcal/mole and more negative values correspond to higher duplex stability.

Curved DNA is believed to play an important role in many cell processes such as transcription initiation and termination, DNA replication and nucleosome positioning [9]. DNA curvature influences the binding affinity of regulatory proteins while DNA looping can increase the proximity of separated regulatory sites. Curvature is defined as the inverse of the radius of an arc that approximates a given DNA sequence. A value of one corresponds to the degree of curvature seen in nucleosomal DNA (see Figure 1, right).

SIDD is a thermodynamic quantity whose value for any DNA base pair may be defined as the incremental free energy (kcal/mole) required to force that base pair to remain open. Regions having low SIDD energy are strongly destabilized, that is, they have a high propensity to melt under normal physiological conditions. The SIDD value for any particular base pair depends on the local GC content and on the superhelicity (degree of negative super-coiling) of the DNA molecule. However unlike stacking energy, SIDD is not purely a local property but rather depends on the distribution of SIDD throughout the molecule. Calculating SIDD for an entire bacterial DNA molecule is a computationally demanding exercise. Even in a 4 Mbp genome, every base pair potentially affects every other base pair.

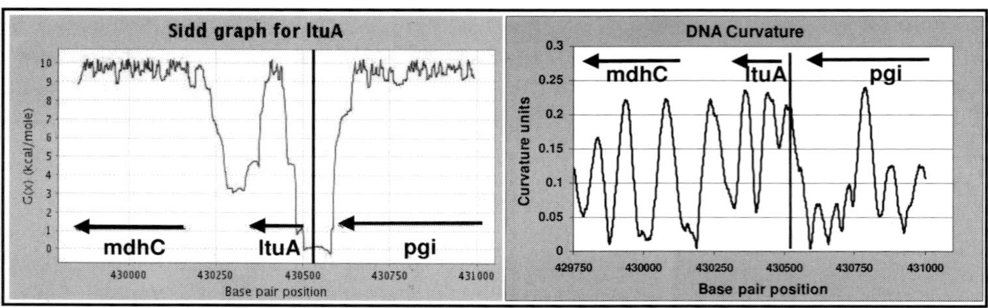

Figure 1. A representation of SIDD (left) and DNA curvature (right) in the vicinity of *ltuA* (*Chlamydia trachomatis*). Vertical lines indicate the TSS location. Horizontal arrows show coding regions. The promoter for *ltuA* lies within a strong SIDD region which occupies the entire upstream non-coding region. It also lies just upstream of a region of high curvature. (Curvature window=100)

Wang and Benham [19] have demonstrated that SIDD energy is a useful predictor of promoter regions. Their reported accuracy of around 80% is, on the face of it, a remarkable result given that the best typical result for promoter prediction in *E. coli* is around 50% [4, 11]. The authors attribute their success to the fact that about 80% of documented promoters contain a strong SIDD site. They define a promoter as extending from positions -80 to +20 with respect to the TSS and they define strong SIDD as any value below 6 kcal/mole. The association of strong SIDD with intergenic regions (see Figure 1, left, for an example) appears to be a general property of all bacterial species [2, 22] although specific association with promoters has been shown only for *E. coli* and *B. subtilis* [19], species for which there is a large number of mapped promoters.

Our group has an interest in the prediction of bacterial promoters using both PWMs and machine learning methods. We have previously shown that the success rates reported for the promoter prediction task are acutely sensitive to the task definition. In particular, the choice of negative instances for the binary prediction task can make the task artificially easy [5] and the degree of focus on regions where TSSs are likely to be found can also bias performance [4].

In this paper we examine the use of DNA structural properties as predictive attributes for finding promoters. Once again we note the sensitivity of the results to the task definition. Our results appear to be the best yet reported for a biologically realistic promoter prediction task. Perhaps surprisingly, structural properties did not contribute appreciably to achieving this result even though they are indeed important for the regulatory activity of many promoters. We also demonstrate for the first time that sigma-54 promoters have a stronger association with SIDD than do promoters associated with sigma-70 and other sigma factors.

2. Methods

2.1. *Data*

All investigations were performed with the genome of Escherichia coli K-12 MG1655 (ACCN:U00096.2) [23]. Experimentally confirmed TSS locations for this genome were obtained from RegulonDB [24]. The data set was filtered for known TSS locations associated with sigma-70 promoters, resulting in 542 records. We extracted 250 bp sequences upstream of those genes closest to the given TSSs. Following Huerta *et al* [7], this approach eliminated all TSSs further than 250 bp from the gene start. We also eliminated seven TSSs that were located within 10 bp of another known TSS because our approach did not discriminate two TSSs closer than 10 bp. The final data set consisted of 439 sequences each 250 bp long, containing a total of 487 annotated TSS locations. Thirty nine of the sequences contained multiple TSS locations.

Stacking energy values were obtained from [1]. SIDD data for the *E. coli* genome were kindly provided by Dr Craig Benham and are available at [22]. DNA curvature was calculated using the CURVA software kindly provided by Dr. Alexander Bolshoy [9].

2.2. *Experimental design*

We approached the promoter prediction task in two steps. First, we constructed a description of a sigma-70 promoter using BioPatML, an XML language for the description of biological patterns [10]. See the next section for more detail of our promoter definition. We scanned all 439 upstream sequences and assigned a score to each position indicating the similarity of that region to our promoter definition as if that position were a TSS. After smoothing the resultant similarity profile with a moving average filter (window = 3), peaks were identified as described in [17] and marked as candidate TSS locations. The rationale is that true TSSs are most likely to be found close to high scoring peaks, that is, locations where the upstream region has high similarity to our promoter definition.

The second step involved using a suitably trained decision tree to classify as *true* or *false* each of the TSS candidates found in step 1. For the decision tree we used the popular WEKA data mining tool [20] and its implementation of C4.5 [14]. This classifier was trained using a selection of promoter features such as the similarity score of the candidate TSS, the -10 and -35 scores of the candidate and a variety of DNA structural features as described later in the paper.

To estimate prediction accuracy, we adopted a 10 fold cross-validation protocol as follows: The 439 sequences were divided into 10 sets. For each fold, nine sets were scanned with our promoter definition to obtain TSS candidates for C4.5 training data. Features were extracted from true TSS locations in each set to obtain positive training instances and from false step 1 TSS candidates to obtain negative training instances.

Testing was performed on candidate predictions obtained from the tenth (holdout) set of sequences. A true positive (TP) was any positive prediction five bp or less from a known TSS. A false positive (FP) was any positive prediction more than five bp from a known TSS. Recall was defined as TP/(TP+FN) and precision as TP/(TP+FP) where FP denotes false positive and FN denotes false negative. Averages were obtained for recall and precision over 10 repeats of 10 fold cross-validation, that is, over 100 folds.

2.3. *Step 1: Use of BioPatML to obtain candidate TSS locations*

Our promoter definition included five elements: an UP element, the -35 element, a spacer, the -10 element and the discriminator (the region between the -10 element and the TSS). The -35 and -10 elements were defined using PWMs prepared from sequence data for known -10 and -35 binding sites available at DPInteract [25]. Scores for the spacer and discriminator widths were calculated using the *accessibility* formula of Shultzberger *et al.* [16, Eq.(2)]. The UP element was defined as a 17 bp sequence, $W_{15}N_2$, directly upstream of the -35 element, where W = A or T and N = any base [11]. Adding an extended -10 element or constraining the TSS to be a Purine did not improve performance.

BioPatML normalises the match score for each pattern element to a value between 0 and 1 - 0 for the minimum possible score and 1 for the maximum. The combined match score is a weighted sum of the normalised element match scores and hence optimisation of the weighting parameters is required. We did this by line search, fixing the weight for the -10 element at 1.0. Interestingly, the optimum weight obtained for the UP element, 0.45, was slightly greater than that for the -35 element, 0.35, indicating the importance of the UP element in *E. coli* promoters.

TSS candidates were obtained by identifying peak locations in the graph of similarity scores. Candidate selection was constrained such that no two candidates could be within 5 bp of each other, i.e. the permitted error tolerance for a correct prediction.

2.4. *Step 2: Classification of candidate TSS locations*

The TSS candidates obtained from step 1 were labeled as *positive* or *negative* depending on their distance from the nearest true TSS. C4.5 training data included the negative candidates from step 1 and positive instances obtained directly from the set of known TSSs. Consequently the available training data consisted of 487 known TSSs (positive class) and 4751 candidate TSSs not biologically confirmed as promoters (negative class)[a].

---

[a] It is likely that some of these negative instances are indeed as yet unidentified promoters.

This over-representation of negative instances was found to reduce the accuracy of the resultant classifier. Consequently, we trained C4.5 with all the positive instances but only the five top ranking (highest scoring) negative instances from each sequence. Note that this set of negatives includes candidates that are most like positives and hence makes the task difficult, albeit realistic. For training, we used the default C4.5 parameters provided by WEKA except that we set the Laplace parameter *true* since this slightly improved performance.

3.   Results and Discussion

3.1 *The TSS Prediction Task*

The first step in our promoter prediction algorithm involved finding candidate TSSs/promoters in each upstream sequence. An average of 11.9 candidates or predictions per sequence was obtained. These were ranked according to their similarity score and each candidate labeled as a TP or FP prediction. Table 1 indicates that of the 217 rank 1 predictions closest to a true TSS, 206 were TP and the remainder FP predictions (>5 bp from the true TSS). The average error of the 217 predictions was 1.98 bp. Recall and precision for the rank 1 predictions were 42% and 47% respectively. This is comparable to the result reported in [11] for the case where TSS-GSS distance was *not* incorporated into the pattern description. Observe that while 90% of true TSSs were within 5 bp of a local maximum, only 42% of them were located within 5 bp of the sequence global maximum. This is consistent with the findings in [7]. As is to be expected, recall increased but precision declined when lower ranked predictions were accepted.

Table 1. Counts of TP TSS predictions obtained from step 1 using a BioPatML description of a promoter. The predictions for each sequence/gene were ranked by BioPatML similarity score.

Rank	# TP predictions	# true TSSs closest to peak	Av error (bp) for predictions
1	206	217	1.98
2	80	92	2.26
3	33	41	2.98
4	33	40	2.88
5	29	32	2.44
6	18	19	2.05
7	12	15	3.47
8	7	9	2.56
9	5	7	2.71
10	7	9	4.44
11	4	4	3.25
12	2	2	1.50
total	436	487	-

The object of step 2 was to design a classifier which could select the true TSS(s) from the candidates identified in step 1. We used the well established C4.5 decision tree because, in our initial investigations, C4.5 outperformed WEKA's implementation of a neural network and an SVM with standard kernels (results not shown).

Success with a classification task depends primarily on identifying appropriate features for the task. Even when a DNA property such as curvature or SIDD is known to play a role in many promoters, finding an appropriate machine learning representation for that feature is not necessarily trivial. We trialed many representations for stacking energy, curvature and SIDD in the vicinity of promoters, the most promising of which are shown in Table 2 along with more obvious features such as TSS-GSS distance.

Table 2. Information Gain merit scores [20] obtained for a range of potential promoter attributes ranked in order of merit. Note that the neighbourhood of a TSS candidate (attributes 4, 10 & 11) is the region -80 to +20 *wrt* the TSS. The promoter upstream region (attributes 8, 9 & 10) refers to -80 to -1 *wrt* the TSS.

Attribute ID	Attribute description	Merit Score
1	Rank of candidate TSS at Step 1.	0.113
2	Distance of candidate TSS from GSS.	0.072
3	Match score of candidate -10 element at Step 1.	0.068
4	Av. similarity score in neighbourhood of candidate TSS at Step 1.	0.066
5	Combined similarity score of candidate TSS at Step 1.	0.061
6	Match score of candidate -35 element at Step 1.	0.028
7	Distance of candidate TSS from position of max. curvature.	0.022
8	GC content of the promoter upstream region.	0.015
9	Stacking energy of the promoter upstream region.	0.014
10	Maximum SIDD gradient in neighbourhood of candidate TSS.	0.008
11	Minimum SIDD value in neighbourhood of candidate TSS.	0.005
12	Maximum curvature in the promoter upstream region.	0.003
13	Is candidate TSS located in low SIDD region? (Boolean)	0.002
14	Is candidate TSS located in the lowest intergenic SIDD region?	0.001

WEKA offers a number of statistical tests to evaluate the efficacy of an attribute when used in isolation for a classification task. Table 2 displays the Information Gain merit scores [20] obtained for a range of potential promoter features ranked in order of merit. The first six features include TSS-GSS distance and various similarity scores obtained from step 1 but do not include DNA structural features, which received low merit scores. The best structural DNA feature was distance of the putative TSS from the position of maximum DNA curvature. Of interest is that maximum SIDD gradient in the vicinity of a promoter was a better feature than minimum SIDD value, so confirming an observation that many TSSs are located near the downstream boundary of a SIDD region.

Based upon the merit scores shown in Table 2, we trained a C4.5 classifier using the first six attributes. This classifier achieved a recall of 50.6% and a precision of 54.0% (Table 3, row 2) on the Step Two task. This was a significant improvement over the results obtained in Step One using the BioPatML promoter description alone (Table 3, row 1). However most of the structural DNA features (such as 7, 8, 9 in Table 2), when added to the basic six features, degraded classification performance. We found two SIDD features and one curvature feature that slightly increased performance when added to the basic six (Table 3, rows 3, 4, 5) but the increase was not significant.

Table 3. Recall and precision for the promoter prediction task obtained after step 1 (selecting promoter candidates) and step 2 (classification of candidates). Feature ID numbers refer to those used in Table 2. Averages are over 10 repeats of 10 fold cross-validation. 95% confidence intervals are shown for the output from step 2.

Step	Feature representation	Include DNA structural features?	Recall	Precision
1	BioPatML	No	42.3%	46.9%
2	1-6	No	50.6±1.6%	54.0±1.4%
2	1-6, 12	Yes	51.1±1.6%	53.9±1.3%
2	1-6, 13, 14	Yes	51.5±1.4%	54.5±1.4%
2	1-6, 12, 13, 14	Yes	51.4±1.4%	54.8±1.4%

A comparison of our best result with previously published results for the TSS prediction task is not straight forward. The difficulty is that there is no standard promoter prediction task and results are sensitive to task definition and the constraints applied. First we must address the issue of task definition. The 80% accuracy achieved by Wang and Benham [19] is readily explained by their task definition. Their positive instances consisted of 500 known promoter sequences which were considered against a set of negative sequences obtained from 500 coding regions and a further 500 convergent non-coding regions, thus yielding a positive-negative ratio of 1:2. The task was then to classify sequences as containing a TSS or not. We have previously argued [5] that this is not the real promoter prediction task because promoters are seldom found in coding sequences or in convergent non-coding regions. The promoter prediction task as addressed in [3, 4, 7 and 11] is to determine the location of promoters/TSSs in regions upstream of gene start sites, since this is where the great majority of promoters are to be found. It is also a much more difficult task because the prediction algorithm must sift through many strong candidates, the majority of which prove to be false instances.

Even where authors agree on the task definition, interpretation is clouded by varying task constraints. Three factors in particular are relevant: (1) the definition of a true positive, (2) the length of the searched upstream region and (3) explicit use (or otherwise) of the TSS-GSS distance. With regard to (1), typically a true positive is a predicted TSS five bp or less from a true TSS. This margin of error is considered acceptable because biological confirmation of an *in silico* prediction does not require greater accuracy. Obviously if the error threshold is tightened, the task becomes more difficult. With regard to (2), 91% of confirmed *E. coli* sigma-70 TSSs are located within 250 bp of the GSS and consequently, most investigations restrict their search to this region. Increasing the distance to 500 bp or more increases the task difficulty as it increases the opportunity to make false positive errors. The search distance also influences the relative performance of algorithms. Using TSS-GSS distance alone as a predictor compares favorably with PWMs where the search distance is 750 bp but not if it is 250 bp [4]. With regard to (3) it has already been noted that prediction accuracy can be increased using TSS-GSS distance because most TSSs are located in a region 30 to 60 bp upstream of the TSS. Whether one considers TSS-GSS distance a valid attribute for this task depends on whether one's goal is to model the behaviour of the RNA polymerase holoenzyme or to find promoters by any means possible.

Table 4. A comparison of the task constraints, recall and precision for several studies of the well defined TSS prediction task.

Authors	Method	Recall	Precision	Error threshold	Explicit use of TSS-GSS distance	Search length
Huerta et al [7]	PWMs	50%	33%	5bp	Yes	250bp
Burden et al [3]	Neural network	50%	17%	3bp	Yes	500bp
Gordon et al [4]	SVM	50%	33%	5bp	No	750bp
Maetschke et al [11]	PWMs+EM	48%	48%	5bp	Yes	250bp
This study	PWMs+C4.5	51%	55%	5bp	Yes	250bp

Our recall and precision values for the TSS prediction task (in Table 3) are compared with four previous sets of published results (see Table 4). Huerta *et al.* [7] in 2003 claimed 'the highest predictive capability reported so far' with a recall of 50% at a precision of 33% (these values are derived from Figure 8e in [7]). We regard this as the benchmark result for a standard set of realistic constraints. Burden *et al.* [3] in 2005 using neural nets, obtained a weaker result probably because they set themselves more difficult task constraints. Gordon *et al.* [4] also obtained a recall of 50% at a precision of 33% but since they searched 750 bp upstream, their task was also notably more difficult. Maetschke *et al.* [11] obtained similar recall but at the significantly higher precision of 48% using an expanded promoter description whose parameters were trained using an Expectation Maximization (EM) approach. Our results offer a modest increase in recall and precision over [11] and therefore represent to our knowledge, the best published result for this task and this set of realistic constraints.

3.2 *SIDD and Promoter Type*

Using the information supplied in RegulonDB [24], we determined the promoter boundaries (-80 to +20 wrt TSS) for all biologically mapped sigma-70, sigma-24, sigma-38, sigma-32 and sigma-54 TSSs. For each TSS, we determined the minimum SIDD value inside its promoter boundaries as defined above. The histograms in Figure 2 show, for each promoter type, the relative frequency of promoters having a given minimum SIDD value. Wang and Benham [19] show similar data but as a probability distribution for all 927 mapped TSS locations in RegulonDB [24]. When we group the promoters according to type, we observe a somewhat uniform distribution of SIDD values for all types except sigma-54, which has 57% of its promoters associated with a SIDD value of less than zero. Only one of the 14 mapped sigma-54 promoters has a minimum SIDD value greater than the strong/weak threshold of 6 kcal/mole set in [19]. It is not surprising that sigma-54 promoters require increased upstream duplex destabilisation. Transcription initiation with sigma-54 requires activation by an enhancer binding protein which binds upstream of the promoter and resulting in interaction with sigma-54 mediated by DNA bending [21].

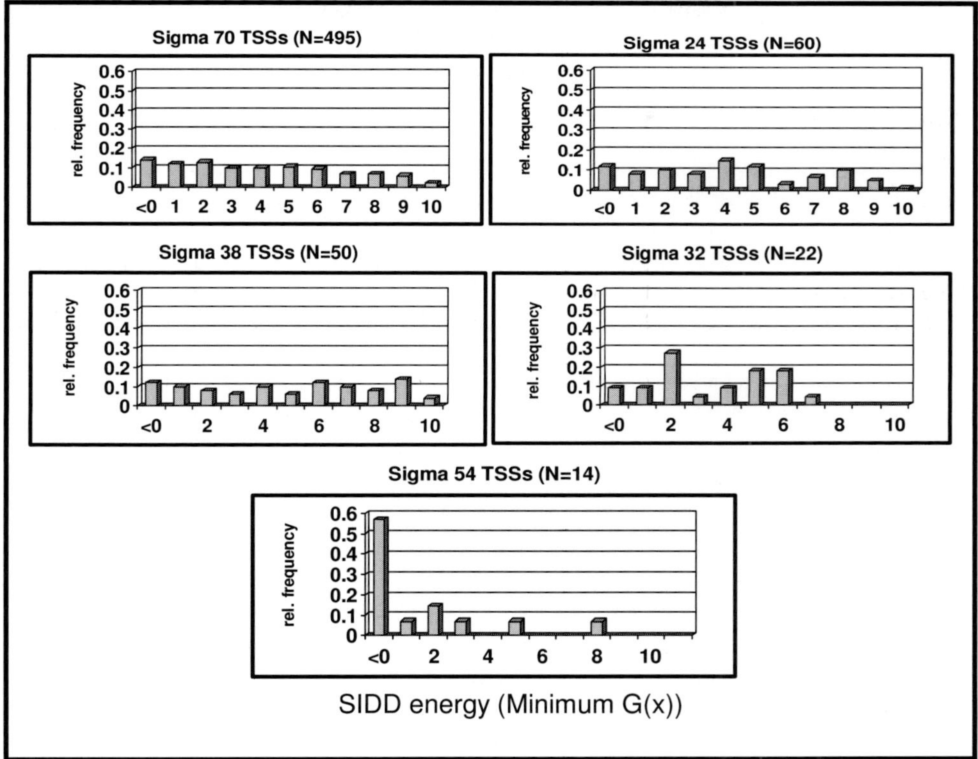

Figure 2. Histograms of the minimum SIDD value associated with five different types of promoter in *E. coli*.

4.   Conclusion

We report in this work the best results to date for a well defined and realistic TSS prediction task. We have also demonstrated that sigma-54 promoters have a stronger association with SIDD regions than do other promoter types. Although DNA structural properties are known to be important in the regulation of many sigma-70 promoters, we were not able to find a suitable representation of these features that helped to increase the *in silico* prediction of sigma-70 promoters. This requires some explanation.

It should be noted that the TSS/promoter task that we undertake in this paper is inherently difficult because it involves the selection of a true TSS from of a set of strong candidates. We find that SIDD regions are generally wide enough to contain several strong candidates and therefore a local SIDD value will not be discriminative. Likewise, regions of high curvature in an intergenic region are sufficiently long and numerous that they do not have strong discriminative value. Finally any selection of promoter features implicitly biases the training towards a particular promoter model but it is known that there are many variations on how promoters initiate transcription, so it is unlikely that any one model or set of features can serve as a general purpose predictor of all the known promoters.

Acknowledgments

This work was supported by the Australian Research Council. The authors would like to acknowledge Dr Craig Benham for providing the *E. coli* genome SIDD data and Dr. Alexander Bolshoy for providing the CURVA software.

References

[1] Baldi, P., Chauvin, Y., Brunak, S., Gorodkin, J. and Pedersen, A., Computational Applications of DNA Structural Analysis, *Proceedings of the Sixth International Conference on Intelligent Systems for Molecular Biology* (ISMB98), 35-42, 1998.

[2] Bi, C. & Benham, C., WebSIDD: Server for Prediction of Stress-induced Duplex Destabilized Sites in Superhelical DNA, *Bioinformatics*, 20, 1477-1479, 2004.

[3] Burden, S., Lin, Y.-X. and Zhang, R. Improving promoter prediction for the NNPP2.2 algorithm: a case study using *Escherichia coli* DNA sequences, *Bioinformatics* 21(5):601-607, 2005

[4] Gordon, J., Towsey, M., Hogan, J., Mathews, S. and Timms, P., Improved prediction of bacterial transcription start sites, *Bioinformatics* 22(2):142-148, 2006.

[5] Gordon, J. and Towsey, M. SVM based prediction of bacterial transcription start sites, *Proceedings 6th International Conference on Intelligent Data Engineering and Automated Learning (IDEAL'05)*, Brisbane, Australia, July 2005. *Lecture Notes in Computer Science*, **3578**:448-453, Springer, Berlin, 2005.

[6] Greishaber, N., Sager, J., Dooley, C., Hayes, S. and Hackstadt, T., Regulation of the *Chlamydia trachomatis* Histone H1-Like Protein Hc2 is IspE Dependent and IhtA Independent, *J. Bact.*, 88(14): 5289–5292, 2006.

[7] Huerta, A. and Collado-Vides, J. Sigma-70 promoters in E. coli: specific transcription in dense regions of overlapping promoter-like signals, *J. Mol. Biol.*, **333**:261-278, 2003.

[8] Kawano, M., Storz, G., Rao, B., Rosner, J. and Robert G. Martin, Detection of low level promoter activity within open reading frame sequences of Escherichia coli, *Nucleic Acids Research*, 33(19):6268-6276, 2005.

[9] Kozobay-Avraham, L., Hosid, S. & Bolshoy, A., Involvement of DNA curvature in intergenic regions of prokaryotes, *Nucleic Acids Research*, 34(8):2316-2327, 2006.

[10] Maetschke, S., Towsey, M. and Hogan, J., BioPatML – an XML description language for patterns in biological sequences, Technical Report, http://eprints.qut.edu.au/archives/00006367, 2007.

[11] Maetschke, S., Towsey, M. and Hogan, J., Bacterial promoter modeling and prediction for *E. coli* and *B. subtilis* with Beagle, *Workshop on Intelligent Systems for Bioinformatics (WISB-2006)*, 9-13, 2006.

[12] Mitchell, J., Zheng, D., Busby, S. and Minchin, S., Identification and analysis of 'extended -10' promoters in Escherichia coli, *Nucleic Acids Research* 31(16):4689-4695, 2003.

[13] Perez-Martin, J., Rojo, F. and de Lorenzo, V., Promoters Responsive to DNA Bending: a Common Theme in Prokaryotic Gene Expression, *Microbiological Reviews*, 58(2):268-290, 1994.

[14] Quinlan, J., *C4.5: Programs for machine learning*, San Francisco: Morgan Kaufmann, 1993.

[15] Schneider, T., Stormo, G., Gold, L. and Ehrenfeucht, A., Information content of binding sites on nucleotide sequences, J. Mol. Biol. 188(3):415-431, 1986.

[16] Shultzberger, R., Chen, Z., Lewis, K. and Schneider, T., Anatomy of Escherichia coli σ^{70} promoters, *Nucleic Acids Research* 35(3):771-788, 2007.

[17] Towsey, M., Gordon, J. and Hogan, J. The Prediction of Bacterial Transcription Start Sites using Support Vector Machines, *International Journal of Neural Systems* **16**(5):363-370, 2006.

[18] Typas, A. and Hengge, R. Differential ability of sigma(s) and sigma70 of *Escherichia coli* to utilize promoters containing half or full up-element sites, *Mol. Microbiol*, 55(1):250-260, 2005.

[19] Wang, H. and Benham, C., Promoter prediction and Annotation of Microbial Genomes Based on DNA Sequence and Structural Responses to Superhelical Stress, *BMC Bioinformatics*, 7:248, 2006.

[20] Witten, I. and Frank, E., *Data Mining: Practical machine learning tools and techniques*, 2nd Edition, Morgan Kaufmann, San Francisco, 2005.

[21] Xu, H. and Hoover, T., Transcriptional regulation at a distance in bacteria. *Current Opinion in Microbiology*, 4:138-144, 2001.

[22] http://www.genomecenter.ucdavis.edu/benham/sidd/index.php

[23] ftp://ftp.ncbi.nih.gov/genbank/genomes/Bacteria/Escherichia_coli_K12/U00096.gbk

[24] http://regulondb.ccg.unam.mx/data/PromoterSet.txt

[25] http://arep.med.harvard.edu/ecoli_matrices/

PART B

Keynote Addresses

DISCOVERING BIOMOLECULAR MECHANISMS WITH PROTEIN SEQUENCE STUDIES: THE ANNOTATOR SOFTWARE SUITE

FRANK EISENHABER

`franke@bii.a-star.edu.sg`

Bioinformatics Institute, 30 Biopolis Street #07-01, Matrix, Singapore 138671

Abstract

Even when it is acknowledge that biomedical sciences are still essentially experimental and they lack a predictive theory in most subfields, it is the more important to underline the few niches where theoretical/computational approaches add creatively to the biological insight. Protein sequence analysis can predict aspects of molecular and cellular function in many cases and, in this way, decisively direct follow-up experiments for the characterization of yet uncharacterized genes and the discovery of new cellular pathways.

Therefore, the analysis of gene/protein sequences is advised to become an integral part of any molecular and cellular biological research, best in the early and planning phase since this allows avoiding unnecessary experiments. At the same time, typical mutational, expression profiling or interaction screens generate dozens or hundreds of protein targets that might require in-depth sequence analysis that, for example with available WWW-tools, will take days for a single target. The ANNOTATOR software suite provides the environment to carry out all routine steps for protein sequence analysis automatically and to enable the researcher to focus her/his time on thinking over the results. The ANNOTATOR has ca. 40 academic tools for protein sequence studies and all major databases built-in together with a number of sophisticated workflows that have shown their potential in previous discoveries.

The talk will give an insight into the biological and software design concepts of the ANNOTATOR. The application discussed include the discoveries SET domain, ATGL and Eco1 functions, the prediction of various posttranslational modifications from protein sequence as well as the extension of the ANNOTATOR for protein mass-spectrometry data analysis tasks.

THE P53 PATHWAY

DAVID LANE

d.p.lane@imcb.a-star.edu.sg

Institute of Molecular and Cell Biology, 61 Biopolis Drive, Proteos, Singapore 138673

Abstract

Somatic mutations in the p53 gene occur in half of all human cancers and germ line mutations in p53 are responsible for the family cancer predisposition known as Li-Fraumeni Syndrome. In those cancers that retain the normal p53 gene other components of the p53 pathway are often damaged. Recently two mouse models have suggested that p53 activity may also affect aging. The p53 response is induced by a wide variety of different stress signals and when activated can induce cell cycle arrest, cell senescence or cell death. Many currently used cancer treatments activate the p53 response through a DNA damage dependant pathway, and p53 gene therapy has recently gained clinical approval in China . In mice and men the threshold of the response is very finally balanced and controlled by a number of regulatory proteins. Of particular interest is the Mdm2 protein, a ubiquitin E3 ligase that binds to p53 and targets it for degradation. A recently discovered polymorphism in the Mdm2 promoter may affect the age of onset of cancer in man. Drugs that target the Mdm2 pathway can act as non-genotoxic activators of the p53 response and one of these is currently in clinical trial in Singapore.

Understanding in detail how the p53 response is regulated may allow the pharmaceutical manipulation of the pathway. We have very recently discovered that the p53 gene has a more complex structure than has been appreciated for the last twenty years and several new iso-forms of p53 have been characterized potentially yielding new sources of individual variation and new targets for therapy.

REGULATION OF GENE EXPRESSION BY SMALL NON-CODING RNAS

HANAH MARGALIT

`hanah@md.huji.ac.il`

The Hebrew University of Jerusalem, Israel

Abstract

Small non-coding RNAs have gained recently much interest, as it has become evident that they are wide-spread in both pro- and eukaryotes and play important roles in post-transcriptional regulation of gene expression. These molecules present intriguing computational challenges: How can small RNA-encoding genes be identified based on the genome sequence? How many such genes are present in a genome? How can their gene targets be identified? What are the properties of regulation by small RNAs in comparison to other types of regulation, such as transcriptional regulation and protein-protein interaction? How is post-transcriptional regulation by small RNAs integrated with transcriptional regulation in the cellular networks? In my talk I will touch upon these questions and describe our attempts to address them. Intriguingly, viruses also encode regulatory RNAs, some of which play a role in cross-talk with the host. By a combination of computational and experimental approaches we identified human targets of viral microRNAs and showed that viruses use microRNAs for evasion of the host immune system. In my talk I will elaborate on these and other interesting human targets of viral microRNAs. Our results have promising therapeutic applications for both immunosuppressive therapy by mimicking the role of the viral microRNAs and for anti-viral therapy by using anti-sense molecules against them.

MAPPING THE TRANSCRIPTIONAL NETWORK IN STEM CELLS REGULATED BY REST

LAWRENCE W. STANTON

stantonl@gis.a-star.edu.sg

Genome Institute of Singapore, 60 Biopolis Street, #02-01, Genome, Singapore 138672

Abstract

REST (RE1 silencing transcription factor) is a protein that regulates neuronal gene expression. REST binds to a highly conserved 21-bp RE1 element and recruits co-repressors to repress transcription. Recently REST was found to be expressed in embryonic stem cell (ESC) and was identified as a direct target of Nanog and Oct4, two transcription factors critical in maintaining the pluripotency and self-renewal of ESC. We are interested in understanding the role that REST plays in ESC. We have identified hundreds of targets genes directly regulated by REST in ESC by performing comprehensive chromatin immunoprecipitation (ChIP)-on-chip experiments. A computational approach was used to identify ~900 RE1 elements within the mouse genome. We then constructed an oligonucleotide array that contained unique probes for all these RE1 sites for our ChIPon-chip experiments. Our results showed that REST binds to > 500 RE1 elements. We are now assessing REST occupancy by a comprehensive sequencing based ChIP method (ChIP-PET) for a more unbiased search for REST targets. Using these two different methods, we are able to identify REST targets at the genome-wide level, which will provide us with a clearer picture of the role of REST in the regulatory network that controls ESC differentiation.

COMPUTATIONAL DISSECTION OF MAMMALIAN REGULATION NETWORKS

MICHAEL ZHANG

mzhang@cshl.edu

Cold Spring Harbor Laboratory , USA

Abstract

Identification of direct targets of an individual or a combination of transcription factors (TFs) is central to determination of regulatory network architecture. Experimental approaches require a combination of expression profiling and binding assay to accurately identify direct targets.

Here we propose an adaptive determination of the gene activation thresholds by using regression splines. Since the thresholds are learnt adaptively from the expression data, the identified targets depend on the physiological condition under which the mRNA sample was obtained. It can work with data from a single condition and no separation of genes into foreground and background sets is necessary. Using human cell-cycle as an example, we show that the E2F targets that we identify at the G1/S phase are significantly different from those at the G2/M phase. We verify known targets and find several novel direct targets of E2F in the G2/M phase.

AUTHOR INDEX